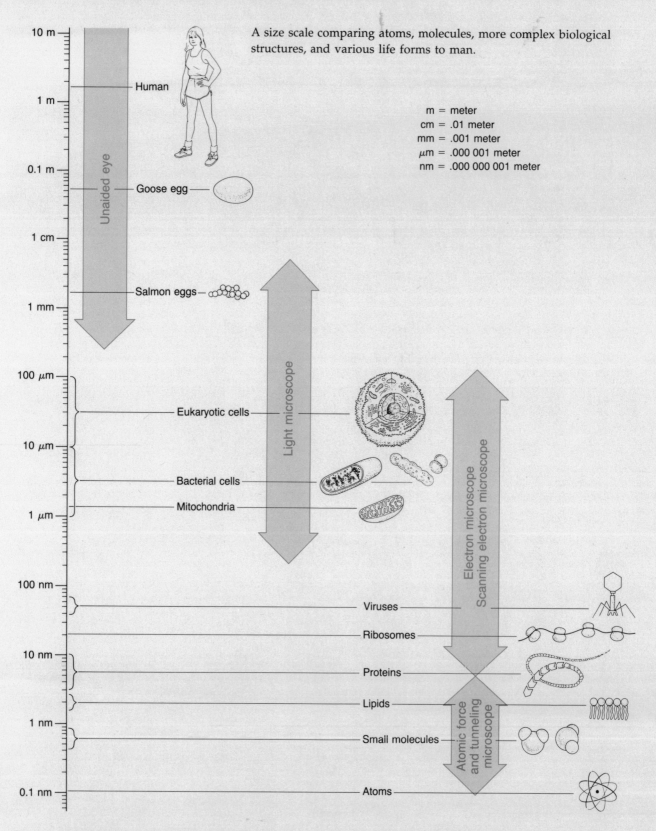

A size scale comparing atoms, molecules, more complex biological structures, and various life forms to man.

m = meter
cm = .01 meter
mm = .001 meter
μm = .000 001 meter
nm = .000 000 001 meter

10 m

1 m — Human

0.1 m — Goose egg

1 cm

1 mm — Salmon eggs

100 μm

Eukaryotic cells

10 μm

Bacterial cells

Mitochondria

1 μm

Unaided eye

Light microscope

Electron microscope
Scanning electron microscope

100 nm — Viruses

Ribosomes

10 nm — Proteins

Lipids

1 nm — Small molecules

Atomic force and tunneling microscope

0.1 nm — Atoms

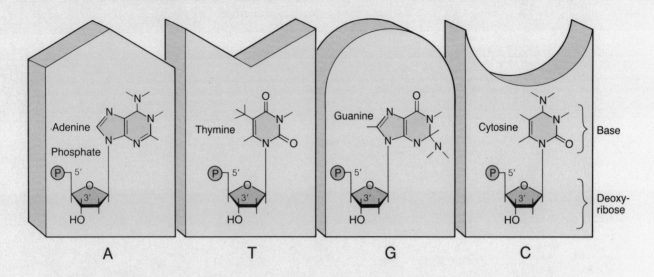

Nucleotides, the basic structural units in DNA. Note that a carbon atom is present (though not shown) at each point where two or more chemical bonds (straight lines) intersect. Bonds with no designated atom at one end are understood to be joined to a hydrogen atom. *(See pages 38 and 39 for the text discussion of this illustration.)*

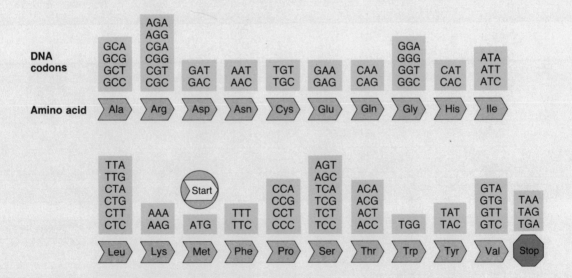

The genetic code. The codons shown for each amino acid are those for DNA. For RNA, the Ts are replaced by Us. *(See pages 63 and 64 for the text discussion of this illustration.)*

Dealing with Genes
THE LANGUAGE OF HEREDITY

Head and Shape 1 (1964)
by Nathan Oliveira

Dealing with Genes
THE LANGUAGE OF HEREDITY

PAUL BERG

Willson Professor of Biochemistry
Director, Beckman Center for Molecular and Genetic Medicine,
Stanford University School of Medicine

MAXINE SINGER

President, Carnegie Institution of Washington
Scientist Emeritus, National Institutes of Health

UNIVERSITY SCIENCE BOOKS
Mill Valley, California

BLACKWELL SCIENTIFIC PUBLICATIONS

University Science Books
20 Edgehill Road
Mill Valley, CA 94941
Fax: (415) 383-3167

Production manager: *Mary Miller*
Manuscript and copy editor: *Gunder Hefta*
Text and jacket designer: *Robert Ishi*
Illustrator: *Georg Klatt*
Indexer: *Gary Cocks*
Proofreader: *Jan McDearmon*
Photo researcher: *Darcy Lanham*
Compositor: *Syntax International*
Color separator: *Color Response*
Printer and binder: *Maple-Vail Book Manufacturing Group*

Library of Congress Catalog Number: 91-75179

ISBN 0-935702-69-5
ISBN 0-632-03339-8 (Blackwell Scientific Publications)

Distributed outside North America by Blackwell Scientific Publications

Editorial offices:
Osney Mead, Oxford OX2 OEL
23 Ainslie Place, Edinburgh EH3 6AJ
54 University Street, Carlton, Victoria 3053, Australia

Orders should be directed as follows:

Marston Book Services Ltd	Australia
PO Box 87	Blackwell Scientific Publications
Oxford OX2 0DT	(Australia) Pty Ltd
(*Orders:* Tel: 865 791156	54 University Street
Fax: 865 791927	Carlton, Victoria 3053
Telex: 837515)	(*Orders:* Tel: 03 347-0300)

To those who shared in these adventures
and in the joys of discovery

Contents

With every advance from the known to the unknown,
the mystery increases.

— LAWRENCE DURRELL
Mountolive

Preface

THIS BOOK has been a long time in the works. The idea for it originated with a series of public lectures given (by P.B.) in 1979 at the University of Pittsburgh. Initial efforts to prepare the lectures for publication were frustrating. Because most of the audience were nonbiologists, many interesting and exciting details of the science were omitted from the lectures, and their absence seemed even more serious in the attempts to prepare written versions of the lectures. The concept for the book began to change, and the need to share the effort of writing became clear. Thus, the two of us began what finally grew into a substantial textbook—called *Genes & Genomes*—for students of molecular genetics. However, throughout the textbook project, we often discussed the original plan and the importance of telling the story to a broader group. Reviewers of the textbook manuscript agreed, and they suggested that the "Perspective" essays that introduce its four sections could serve as the basis for a book for general readers.

So, with their helpful and encouraging comments in hand, and *Genes & Genomes* complete, we began writing this book. We are grateful to those reviewers for encouraging us to share with the public that supports the research, the science that will affect all our lives in the future. We especially want to express our appreciation to Bruce Armbruster, President of University Science Books, for encouraging us to undertake this project and for his unfailing support during our labors.

Much of the work on this book, at all stages, was aimed at making the ideas of modern genetics accessible to a broad audience. Biology and genetics, even molecular genetics, in contrast to modern physics, do not much depend on abstract concepts, nor does their understanding require a mastery of mathematics. Yet, although it is not abstract, biology is complex. The variety of living things we all encounter daily—viruses, bacteria, plants, insects, and mammals, for example—reveals biology's complexity, as do the workings of our own bodies. The complexity and novelty of biological mechanisms require a special vocabulary. Moreover, an understanding of molecular genetics, by its very nature, depends on some familiarity with chemistry. It has been our goal to minimize these hurdles for the general reader, but at the same time, not to trivialize the extra-ordinary depth and breadth of biologists' current understanding of life processes. We hope that readers will stay with us through some of the tougher parts. Their reward will be a sense of the relevance and beauty of some of the most fascinating recent discoveries about nature. Most of the more difficult material is in the first four chapters, where important fundamental ideas, the chemistry of biological molecules, the basic chemical processes of heredity, and most of the new words are introduced. Thereafter, at least according to the reviewers, the material becomes a good deal easier.

Only very rarely do we mention the names of the scientists responsible for major conceptual or technical advances. This is the result of a conscious decision. It may strike some readers as odd and other readers as unfair, and it may detract from the pleasure of those who, with good reason, think that the study of science is enhanced by knowing about the discoverers and how their idiosyncratic minds permitted great intellectual leaps. Therefore, we apologize to those who will miss the personal details. The fact is that, with rare exceptions, the scientific breakthroughs described here depended on the work of many investigators and several "generations" of scientists. Sorting out proper credit can be a difficult task, and the chance of slighting colleagues is real. Besides, we believe it is the science itself and the extraordinary consequent insights into nature that should hold center stage.

We and most of our colleagues think about genetics visually. We "see" genes and the machinery that transforms their information into traits in our minds' eyes. The diagrams in scientific papers are critical to our understanding and our memory of intricate details. Therefore, we remind readers that the drawings prepared for this book are an in-tegral part of the narrative. They clarify many points in the text and also imply more than we have been able to describe in words. We are

PREFACE   xiii

especially grateful to Georg Klatt and his collaborator, Audre Newman, for transforming our rough sketches into artistic renderings of the ideas we meant to convey. Georg's devotion to this project and his creative talents exceeded even our fondest hopes for this important feature of the book. Bob Ishi's fine sense of design, particularly of the cover, provides a hospitable format for the book's contents. Our gratitude to Nathan Oliveira for allowing us to reproduce one of his many wonderful art works as the frontispiece is boundless. Its imagery captures the sense of mystery we see in biology.

Two nonscientists, Robert L. Stern and Daniel M. Singer, struggled through an early draft of this book, pointing out confusing jargon and hidden assumptions. A revised version was read critically by Hugh A. D'Andrade, Peter Armbruster, Steve Olson, John D. Roberts, David B. Singer, Frank M. Turner, and Michael P. Yaffe, and we are deeply grateful for their insights and suggestions. Additional revisions were made, and the third manuscript version was then read by Barbara Bowman, Christopher Chyba and Russell F. Doolittle. Their comments helped us to greatly improve the clarity of the story. We are particularly grateful to Douglas R. Hofstadter for his extraordinarily perceptive and provocative review of the manuscript, and for the clarifications he suggested. Throughout, Carol Dempster did a masterful job of correlating all the comments and helping us to edit and revise the text accordingly. The final version is enormously enhanced by her interest, wisdom, and good humor. Gunder Hefta put the finishing touches to what we believed was the final version, an enhancement we came to appreciate greatly. The errors that remain in spite of all this good help are entirely our own.

Among the splendid advantages available to modern biologists are the scientific libraries wisely established by earlier generations of scientists. During our work, we have spent a good deal of time reading and writing at three such places, all conveniently remote from telephones and daily responsibilities. The librarians and other staff members of the Marine Biological Laboratory at Woods Hole, Massachusetts, of the Jackson Laboratory at Bar Harbor, Maine, and of the Stanford University Hopkins Marine Station at Pacific Grove, California, were all hospitable and helpful.

Harry Woolf, formerly Director of the Institute for Advanced Studies at Princeton, graciously provided us with the opportunity to work at that idyllic place. His kind interest, hospitality, and friendship, and the fine food offered by the Institute's eclectic chef, Franz Moehn, made those days memorable. Curiously, our ability to work uninterrupted at Princeton was assured by the almost universal lack of interest on the part of the Institute's physicists and mathematicians in talking to biologists. A second stay at the Institute, sponsored by its current director Phillip Griffiths, allowed us to review the final proofs before publication.

Relatively uninterrupted days for work were also assured when one of us (P.B.) was a Fellow of Clare Hall, Cambridge. The lively intellectual community there, presided over by Michael Stoker, was congenial and stimulating. A fellowship from the John Simon Guggenheim Memorial Foundation similarly afforded one of us (M. S.) an extended period for writing.

Over the years, the science we hoped to capture on these pages progressed at an extraordinary and unprecedented pace. Continual revisions put heavy demands on Eleanor Olsen and Dorothy Potter at Stanford, May Liu and Gail Gray in Bethesda, and Sharon Bassin in Washington, the good-humored and skillful people who typed and retyped the innumerable drafts of this book.

Our spouses, Millie Berg and Dan Singer, have been understanding and patient, and their support was always clear and always welcomed. And the younger members of the Berg and Singer families, though none is a biochemist or molecular biologist, have bolstered us with their curiosity, enthusiasm, pride, and love.

Paul Berg
Maxine Singer

Dealing with Genes
THE LANGUAGE OF HEREDITY

…the search for simplicity in the form and function of biological organisms has been most fruitful when it has led to the discovery of new complexity.

— JOSEPH S. FRUTON
A Skeptical Biochemist

Prologue

GENETICS, once an arcane branch of biology, has become front-page news. The business sections of daily newspapers regularly report the ups and downs of the biotechnology industry. Medical reporters describe the latest experiments having but the remotest possible significance to disease diagnosis and treatment. The supporters of the Human Genome Project are asking for major new expenditures of federal funds. Economic and scientific pundits see in the new genetics an answer to economic woes and a rebirth of the United States' competitive advantage. Environmentalists, religious figures, sociologists, and philosophers are spending their time and energy considering the implications of genetics. Government regulators worry about how to control the materials and organisms produced by new genetic techniques. Patent attorneys and judges are pondering the question of which organisms should be patentable.

Underlying all this tangential activity are extraordinary scientific findings, findings that place our understanding of the mechanisms of inheritance at the molecular level. No longer does our understanding of how heredity works depend entirely on breeding experiments and the chance identification of mutant traits. No longer is the unit of inheritance, the gene, merely an abstract notion. Genes and the chromosomes of which they are a part are now describable in precise chemical terms. Even more significantly, genes can be synthesized, manipulated and altered by chemical methods in the laboratory and reintroduced into the cells of living organisms.

The observations of a nineteenth-century monk, Gregor Mendel, on the inheritance of simple traits, such as the color of peas, has developed into a profound modern science. The story of these developments inspires biologists of all kinds because it carries with it the promise that the living natural world is not only diverse and beautiful but also understandable. The story can certainly be appreciated by nonbiologists. It can enhance anyone's understanding of the biotechnology industry, the developing medical, industrial, and agricultural implications of genetics and genetic engineering, and of the significance of the international Human Genome Project.

1

Genetics is the scientific study of inheritance. Its roots lie in the earliest recognition that like begets like—but not always: Sometimes an offspring is curiously different from its parents. Early humans discovered that they could exploit such changes in useful plants or animals. Deleterious differences could be avoided and advantageous ones preserved and improved by selective breeding. Moreover, such curious differences appeared in humans as well as in other animals and in plants.

Prehistoric humans were good practical geneticists; the proof lies in the domesticated animals, plants, and microbes that feed and clothe and work for us to this day. But scientific genetics began in the nineteenth century with Mendel's careful studies of inheritance in pea plants. From Mendel through the middle of the twentieth century, the study of genetics depended on two things. First, it required an organism that could be bred in a directed way so that, over several generations, the relation between parents and offspring could be known unambiguously. Second, it required some readily identifiable inherited characteristics of the species of organism being studied. Thus, Mendel followed the color of pea seeds (green or yellow) and the height of the pea plant (tall or short). Later, geneticists bred fruit flies and studied the inheritance of eye color (red or white) and wing structure (full-sized or truncated). Where one form of the trait predominated, such as red eyes or full-sized wings in fruit flies, it was considered normal; unusual forms, such as white eyes or truncated wings, were called **mutants**. During this era, genetic research depended on two techniques: experimental breeding and arithmetic (counting the frequency of normal and mutant traits among the offspring). From this work, great and enduring laws of inheritance were deduced, as will be seen in Chapter 1.

For ethical reasons, experimental breeding cannot be carried out with humans. But centuries of accumulated observations suggested that the concepts developed with pea plants and fruit flies were applicable to humans as well. Differences in many common human traits—such as the color of skin, hair, and eyes—were clearly inherited. Less obvious and less common traits, including malformed limbs, colorblindness, and certain diseases (such as hemophilia) ran in families, and their inheritance seemed to follow certain rules. For example, colorblindness and hemophilia occur predominantly in males but are inherited from the mother.

By the middle of the twentieth century, the fundamental rules governing the inheritance of traits had been elucidated; and the rules were the same for plants, for insects, and for mammals. The achievements of this era, of what is now called **classical genetics**, were profound. Yet genetics remained dependent on breeding and counting, and its conclusions were abstractions.

No one could describe what a gene might actually be. People knew only that there were genes for eye color, for wing shape, for flower color, for seed shape, and for the ability of humans to see color or to produce blood that would clot normally. And they knew that any particular gene could occur in several forms, specifying red or white flowers, tall or short plants, hemophilia or normally clotting blood.

Genetics is now a very different science. While breeding and counting are still important, much of genetics has focused on chemistry. The new **molecular genetics** described in this book began to develop in the 1950s. Its origins were in classical genetics, in the distinctive science of microbial genetics (and especially in the study of viruses that infect bacteria), and in biochemistry. Two distinct periods in the development of molecular genetics can be identified. The first began around 1945 and extended into the early 1970s. This period's great achievements include, first, the recognition that, in all living things, genes reside in large molecules called DNA; second, the description of the chemical structure of DNA; and, third, how the DNA structure specifies an individual's traits, such as color, shape, and susceptibility to particular diseases, as well as those functions that enable us to digest food and to use it for growth and energy.

The second period, starting roughly around 1970 and continuing through the present, began with the development of methods that together are called **recombinant DNA techniques** or **genetic engineering**. These revolutionary methods allow the isolation and characterization of individual genes from any organism, be it microbe, plant, or animal. As a result, we now know the differences in gene structure that yield normal color vision or colorblindness. We also know precisely how a gene required for normal blood clotting is altered in individuals afflicted with hemophilia. We know how mutations leading to white eyes in fruit flies disrupt the gene responsible for the flies' normally red eyes. Extensions of the new methods have also led to novel diagnostic tools for inherited human diseases as well as for cancer and AIDS. Similar applications are being used·in forensic medicine for the positive identification of victims and criminals. There is also substantial reason to believe that the new knowledge can lead to innovative and effective therapy for diseases of humans and the development of economically important animals and plants. Because molecular genetics is applicable to all organisms, it also provides new routes to a deeper understanding of the stresses to which our planet's ecosystems are now subjected.

The significance of such dramatic improvements in diagnostic techniques and of the potential for new therapies is enormous. More than 4000 human disorders are known to be caused by defects in one or a few genes. Many of these disorders are debilitating or lethal. Increasingly, medical research is revealing genetic components in other common diseases. Cancer, for instance, is fundamentally a genetic disease that involves changes in DNA in the cancerous tissues. Heart disease and one of its important determinants, blood cholesterol levels, are affected by genes. So too are susceptibilities to rheumatoid arthritis, schizophrenia, manic-depressive psychosis, and juvenile diabetes, to name but a few.

The deeper understanding emerging from current molecular genetics reflects the fact that we can deal directly with the source of the fundamental properties that distinguish living from nonliving things—DNA molecules. But being able to deal with genes means more than simply understanding them. The science of molecular genetics also allows us to alter genes. DNA molecules can be changed, mutations can be introduced or eliminated, and the altered genes can be reintroduced into certain living experimental organisms. Such techniques are currently being applied to bacteria, yeast, some plants, fruit flies, and mammals. As we write, the first attempts are being made to deliver genes into patients suffering from certain genetic disorders or forms of cancer in the hope of developing new and effective therapies for these and other diseases.

Thus, what some pundits have labeled the "revolution in biology" has initiated more than a new era for understanding living things. It has initiated research with unprecedented implications respecting our ability to manipulate living things, including ourselves, in fundamental ways. Human nature being what it is, and conceivably not alterable by genetic means, it is a tremendous challenge to apply these extraordinary abilities for the betterment of our planet and all its living inhabitants. For this reason, the "revolution in biology" has also heightened social concern over the impact of biological research.

As it happened, the earliest public concerns reflected questions raised by the community of biological scientists and centered on the safety of recombinant DNA experiments—experiments in which DNA molecules from unrelated organisms are joined together in the laboratory and propagated in living cells. Laboratory safety has always concerned people working with pathogenic (disease-causing) microbes, such as viruses and bacteria; and these concerns were renewed in the early 1970s as more work was carried out worldwide on viruses that cause tumors in experimental animals. The distinctive life cycles of these viruses (compared with the more familiar viruses that cause, for example, measles and polio), and the possibility that some of them might cause human cancers, prompted a serious look at the possible hazards to scientists, their colleagues, and the public at large. Soon after, the recombinant DNA methods emerged.

The earliest proposal for recombinant DNA experiments called for the joining of DNA from a virus that causes tumors in small laboratory rodents with DNA from a virus that exclusively infects bacteria. Plans to introduce the new DNA molecule into bacterial cells aroused serious reservations among some scientists, who thought the bacteria might then be able to cause tumors in humans. These concerns were stilled when the creators of this recombinant DNA decided to defer their plans. The next major step was the construction of recombinant DNA molecules containing DNA corresponding to a gene that makes bacterial cells resistant to antibiotics. The ability to introduce such novel DNA molecules into living bacteria raised questions about whether such experiments could lead to the creation of bacteria that would be resistant to medically important antibiotics. The issues were raised by scientists at scientific meetings and in private conversations. At one such meeting in June 1973, the participating scientists called attention to this promising line of research and asked the National Academy of Sciences to undertake a focused study of the possible risks of recombinant DNA experimentation. These concerns were publicized in a letter published in the scientific weeklies *Science* and *Nature*.

The Academy responded (as academies typically do) by forming a committee, this one of scientists actively engaged in recombinant DNA research and related work. This committee met in April 1974 and took two major steps that became, to their surprise, front-page news. First, they called for a worldwide moratorium on recombinant DNA experiments using tumor virus DNAs or resulting in the introduction of genes responsible for potent toxins or antibiotic resistance into bacteria that normally do not have such genes. Second, they called for an international, broad-gauged discussion of the issues, to be held at a meeting the following winter.

Although there were rumblings of discontent, and even accusations that U.S. scientists were trying to slow up everyone else's work so that they could win a race to major new discoveries, the moratorium was, as far as anyone knows, universally honored. Scientists from the United States and abroad, as well as science administrators, journalists, and even lawyers, readily accepted invitations to the proposed conference.

The conference took place in February 1975 at Asilomar, a conference center run by the State of California and located on the edge of the Pacific, a few miles south of Stanford University's Hopkins Marine Station in Pacific Grove. The organizing committee had arranged for prior working-group sessions and working-paper preparations by experts in several different areas. There was a good deal of discussion, some of it heated, about the reality of some of the dangers being discussed. There was broad agreement (but not unanimity) that, if the risks of constructing hazardous organisms were real, their likelihood was very low. But consensus on a future course of action seemed remote until the lawyers took their turn. They underscored the personal legal responsibilities of scientists, even for highly unlikely events, and they reminded the gathered scientists that the public can be very restrictive about remote risks if its fear of the consequences, however unlikely, is great. The message was clear: Continuing to be conservative was the only sensible course. On the last day, the organizing committee's proposed report was debated vigorously but finally accepted. The recommendations of the Asilomar conference were widely reported in the press and later published in several scientific journals. Similar conclusions had been reached by a British government committee under Lord Ashby a few weeks earlier.

The recommendations made at Asilomar provided the framework and starting point for official U.S. action, which began the day after the conference closed. A committee organized by the National Institutes of Health (NIH) began work on guidelines to govern all recombinant DNA experiments supported by government funding. The original guidelines, published in June 1976, were intentionally strict, with the expectation that, as experience and knowledge accumulated, they could be revised. To this day, no untoward events affecting laboratory personnel or the public are known to have originated from the hundreds of thousands of recombinant DNA experiments that have been conducted. The strict requirements regarding laboratory design and experimental procedures for most routine recombinant DNA experimentation have now been relaxed or eliminated. The only experiments that still require strict compliance with the NIH guidelines (or their equivalents in most other countries) are those that include recombinants with extensive DNA regions from highly pathogenic organisms. Interestingly, recombinant DNA techniques have made the study of some important but very dangerous infectious agents of humans and animals both feasible and safe.

Since the Asilomar conference, public interest and concern over recombinant DNA experiments have continued with varying intensity until the present. Of particular concern has been the genetic engineering of whole organisms. At first, the public's focus was the same as that of the majority of the scientific community—the potential of the experiments to create disease-producing agents. State and local governments passed laws and ordinances mandating compliance with the NIH guidelines or, in some cases, somewhat more restrictive rules. Bills proposing to make the NIH guidelines law, and to introduce punishments for noncompliance, were introduced and debated in Congress, but none ever passed. These many independent debates, in various locales and at many levels, ultimately validated the NIH approach to the problem and their recommendations for its management.

Later, and to this day, other kinds of issues have dominated public debate. A few scientists raised questions about the possible evolutionary consequences of moving DNA across species boundaries—for example, inserting human DNA into bacteria. These questions attracted attention from many nonscientists, but the issue is now viewed by most scientists as inconsequential. There are myriad opportunities in nature for such exchanges of DNA to occur, and they have probably happened many times; recombinant DNA experiments do not significantly amplify the opportunities. Moreover, most bacteria chosen to harbor recombinant DNA molecules are able to grow only under very special laboratory conditions. Similarly, for various agricultural purposes, government regulations require the use of special genetically engineered organisms that can grow only in restricted environments. Although careful consideration of the potential problems associated with such releases is essential, requirements such as the use of protective garb resembling astronaut suits are, from a scientific standpoint, excessive.

The possible application of recombinant DNA techniques to the development of biological warfare agents is raised continually. The United States is party to a 1972 international convention prohibiting such work, but this is not true of all countries. The U.S. Defense Department continues to support work aimed at defense against biological weapons that might be developed by others. Some people object to this policy because research intended for defensive purposes overlaps with research needed to develop offensive capability. This debate properly attracts scientists and nonscientists alike, because it is fundamentally a political and social question, not a scientific one. So too are discussions by the public of other questions arising out of the new biology. What are the advantages and disadvantages of gene therapy for humans, should it ever become feasible? How should society, employers, and individuals deal with the growing amount of information about people's genes that can be acquired as a result of the new technologies? Are there valid ethical concerns associated with the introduction of human genes into animals? Should genetically engineered animals and plants be patentable? What are the ethical issues that will arise from the Human Genome Project?

The early public fears about the revolution in biology engendered many negative attitudes about the research. Biologists then feared the worst: highly restrictive laws or regulations that would seriously hinder further experimentation and its promise of new knowledge and beneficial applications to medicine, agriculture, and industry. Many scientists regretted the initial open discussion of the issues because of what appeared to be successful demagoguery by their critics and a tendency of newspapers to encourage hype rather than to explore difficult issues carefully.

Yet it has turned out well. The science described in this book attests to that, as does the growing number of important products being produced by the young and energetic biotechnology industry. Perhaps the moratorium and the initially restrictive guidelines held research up, but only briefly. The early caution in the face of ignorance was prudent, even though hindsight suggests that the risk scenarios were far less likely than many scientists had supposed.

CHAPTER 1

Genes Revealed

OUR MOST ancient ancestors must have been curious about heredity. They had only to look at the rich natural life that surrounded them and realize that trees beget trees, birds beget birds, and they themselves beget new human beings—and then to wonder why. We recognize the signs of their curiosity in ancient myths. In a past even more distant than the myths, our ancestors began to do more than wonder. They began to experiment and then to manipulate heredity for their own benefit. We have their legacy in the domesticated plants and animals we still depend upon for our food and clothing, and in yeast adapted for making bread and wine and beer. And in this legacy, we find the origins of the disciplined study (that is, the scientific study) of heredity that came to be called **genetics**.

Historically, the extraordinary diversity of living things was an obstacle to the discovery of unifying principles about biology in general and about heredity in particular. It is not easy, after all, to see the relation between a tree and a horse. Major concepts developed slowly, and some good ideas were forgotten, only to be rediscovered centuries later. Fifth-century Greek philosophers, defying strong cultural pressure to confirm male dominance, concluded what now seems obvious: Because offspring resemble both parents, both sexes must contribute to the formation of a new individual. They also believed that those contributions were information of some kind, collected from parts of mature individuals to form male and female "semen." Democritus, whose view did not prevail, suggested that the information was carried in the form of particles, whose shape, size, and arrangement influenced the properties of the offspring. Theophrastus, a student of Aristotle, was the first to recognize the similarity between animal and plant reproduction, and he proposed that the ideas of "male" and "female" be used to designate the participants in sexual reproduction.

Genetic research began in earnest in the nineteenth century. Inheritance of variable, externally obvious traits of readily available plants and animals, such as the color of pea flowers or of human eyes, was studied. Emerging from these investigations was the abstract concept of an indivisible **gene** as the fundamental unit of inheritance. Thus, it was supposed that one gene would determine the color of the pea flower, another the height of the pea plant, and so on. Insights into the nature of genes themselves or how they might determine traits were nonexistent.

Explorations into the chemical identity of genes and the details of the mechanisms governing heredity began only in the middle of the twentieth century, when fungi, bacteria, and viruses replaced plants and animals as experimental models. It was in these simpler life forms that **deoxyribonucleic acid (DNA)**, **ribonucleic acid (RNA)**, and **proteins** were first recognized as the universal determinants of the traits of living things. Great advances in understanding the hereditary mechanisms used by fungi, bacteria, and viruses followed speedily and decisively because their biological characteristics simplified the analysis of their genetic structures. Implicit in these advances was the concept of the gene as information—that is, information that governs the characteristics, growth, and behavior of living things. A consequence of this was that all living things must have mechanisms that decipher or "read" the genetic information. Genes also must permanently store such information for transmission from parent to offspring.

Because comparable analytic capabilities for examining more complex organisms were lacking, the extension of these insights to animals and plants was delayed. Then, in the early 1970s, the development of revolutionary new techniques erased the formidable technical and conceptual barriers to understanding complex genetic systems. Not surprisingly, our views of the structure, organization, and function of genes were modified profoundly. This new understanding is, in turn, radically altering broad areas of biology. We can obtain some perspective on these advances by examining the origins and successive revisions of fundamental concepts about inheritance.

Cells

EARLY in the nineteenth century, following the introduction of improved microscopes, the concept of the cell provided a major unifying simplification for biology. All organisms live either as single cells multiplying independently in their environment or as coherent collections of cells. Bacteria, for example, are single cells, and generally each cell can function independently of others during growth. By contrast, plants and animals are complex, consisting of thousands, millions, billions, or trillions of cells of different sizes and shapes. In such complex organisms, different kinds of cells have different functions. Often, cells of a particular kind are joined together to form more complex structures, such as the tissues that constitute a liver, or brain, or leaf. Tissues usually carry out specific functions that are determined by the properties of the tissue's cells. One of the most important advances in nineteenth-century biology was the recognition that new cells arise only by the division of preexisting cells.

Some cells—the eggs of fish and frogs, for example—are quite large, but most cells can be seen only with the aid of a microscope. Increasingly sophisticated refinements in microscope optics and innovative methods for staining biological materials with dyes have provided more detailed descriptions of cells and their interiors (Figure 1.1). Although the shapes of cells vary, each may be viewed as an assortment of biological molecules inside a sac bounded by a fatty layer called the **plasma membrane**. The material inside the cell, the **cytoplasm**, contains many distinct and readily identifiable structures. Each of these structures is associated with particular cellular processes. In animal and plant cells, the mitochondria are responsible for the production of energy needed for chemical reactions, locomotion, and growth; the endoplasmic reticulum and the Golgi bodies transport molecules from one place to another within the cellular space or secrete them to the cell's surroundings. Highly specialized cells have characteristic structures that provide for specialized functions. A good example of such structures is provided by the chloroplasts found in the cells of green plants (Figure 1.1). Chloroplasts are the organelles (literally, "small organs") in which energy in the form of sunlight is converted to chemical energy for use by plants and, ultimately, by the animals that feed on plants. Aside from having many structural elements, the cell is full of water, in which are dissolved a large variety of molecules, from simple salts to complex sugars and proteins. The most striking and usually the largest structure inside animal and plant cells is the **nucleus** (plural, *nuclei*), which is itself a sac. The largest structures visible in the nucleus are nucleoli.

Today, living things are divided into two groups, depending on whether or not their cells contain nuclei. Cells that possess nuclei—for example, multicellular organisms such as plants and animals, as well as some unicellular organisms, such as yeast and protozoa—are called **eukaryotes**. The second group, the **prokaryotes**, lack nuclei; the only organisms that fall into this group are the single-celled bacteria (Figure 1.2). [These terms derive from the Greek word *karyon*, meaning "kernel" (nucleus), and either *pro-*, meaning "before," or *eu-*, meaning "good." Thus, eukaryotes have a "good" nucleus,

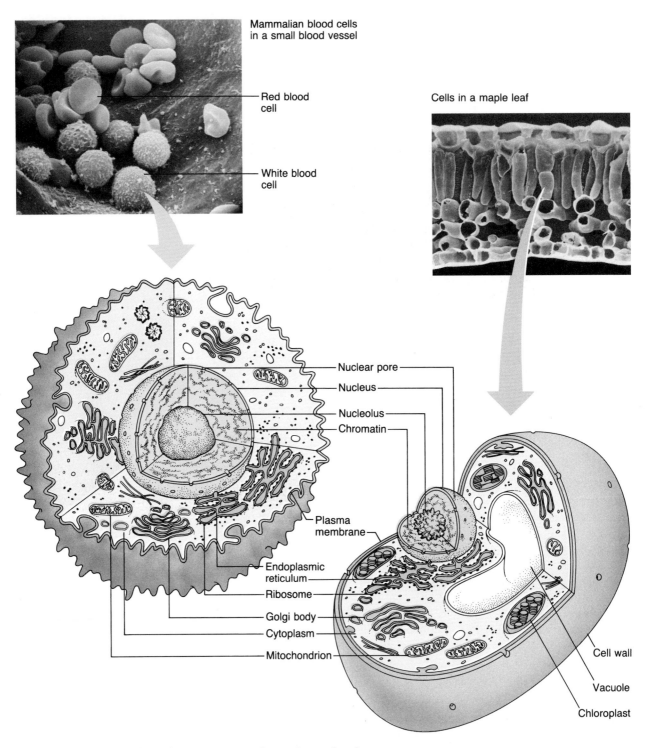

Mammalian blood cells
in a small blood vessel

Red blood
cell

White blood
cell

Cells in a maple leaf

Nuclear pore
Nucleus
Nucleolus
Chromatin

Plasma
membrane

Endoplasmic
reticulum
Ribosome
Golgi body
Cytoplasm
Mitochondrion

Cell wall

Vacuole

Chloroplast

Figure 1.1 Photographs and cutaway views of typical animal and plant cells. Some important internal structures that will be referred to in this book are indicated.

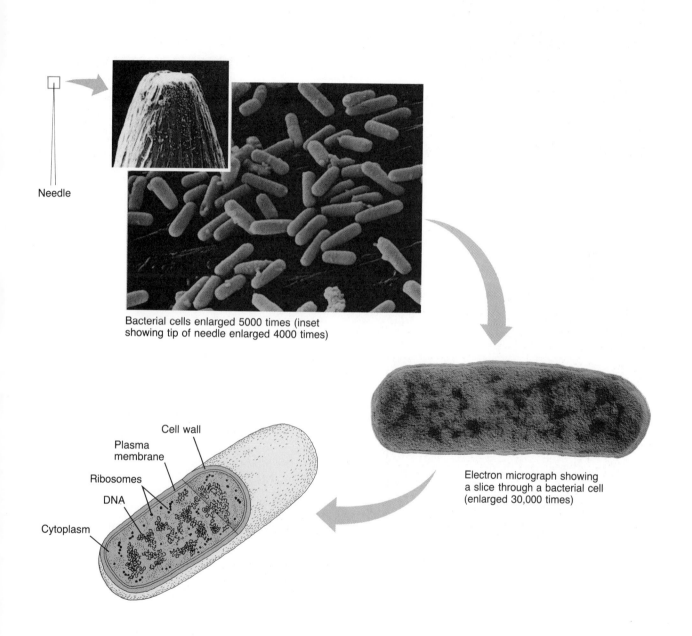

Needle

Bacterial cells enlarged 5000 times (inset
showing tip of needle enlarged 4000 times)

Electron micrograph showing
a slice through a bacterial cell
(enlarged 30,000 times)

Cell wall

Plasma
membrane

Ribosomes

DNA

Cytoplasm

Figure 1.2 Photographs and a cutaway view of a typical bacterial cell.

while prokaryotes do not.] Eukaryotic cells are larger and more complex than prokaryotic cells, and they generally contain more genetic information. Furthermore, eukaryotes are capable of true sexual reproduction; for many, the sexual mode is obligatory for the production of offspring. Viruses, which have their own genes, are not cells at all and cannot multiply independently. Instead, they must infect living cells, where their genes function to produce new viruses, often killing the infected cells in the process.

Soon after the cell theory was defined, three approaches to the study of living things were initiated: the statistical analysis of how single traits are inherited, the study of chromosomes within the nucleus, and the chemical characterization of cellular and chromosomal constituents. These approaches proceeded in parallel and developed into important, independent scientific disciplines before merging in the middle of the present century.

Chromosomes

E ARLY in the nineteenth century, biologists observed small structures inside the nuclei of plant and animal cells. They called these structures **chromosomes** from the Greek *chromo*, "color" and *soma*, "body," because they became especially brightly colored when cells being prepared for microscopic examination were treated with certain dyes. By the second half of the nineteenth century, biologists had acquired a detailed picture of the form and behavior of chromosomes. It was realized that all the cells in any individual of a given eukaryotic species (with the important exception of eggs and sperm) have the same characteristic number of chromosomes. For example, certain fruit flies have eight chromosomes, while humans and evening bats have 46, corn has 20, and rhinoceroses have 84. This is so even though an individual's cells are quite different in structure and function. Figure 1.3 shows human chromosomes as seen by light and electron microscopy. The chromosomes can be grouped into pairs on the basis of similar shapes: four pairs in fruit flies, 23 pairs in humans, and so forth. The two similar members of a pair are said to be **homologous** to one another.

Microscopic examination of chromosomes in dead cells that have been stained gave only static pictures of their behavior. However, it was possible to put such pictures in a sequence, beginning with the formation of two new cells by the division of an old one and ending with the division of

15

Light micrograph of
human chromosomes
(enlarged 600 times)

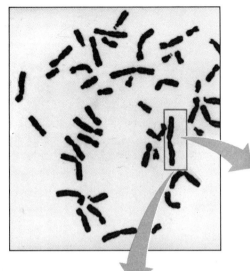

Electron micrographic 3-D image
(enlarged 30,000 times)

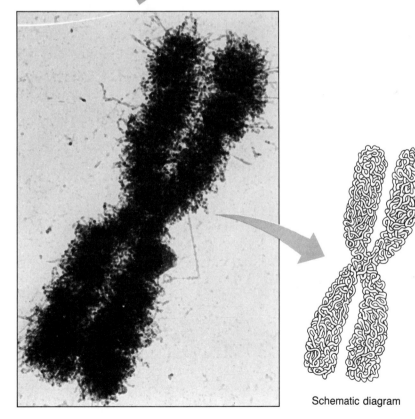

Electron micrograph
of fixed chromosome
(enlarged 30,000 times)

Schematic diagram

Figure 1.3 Human chromosomes.

these two into four. When viewed in order, such pictures showed how chromosomes change during the time a cell prepares to divide, much as serial pictures in a book can illustrate action when the pages are flipped. It became apparent that, in the course of cell division, a duplicate of each chromosome is made, resulting in a doubling of the number of chromosomes (Figure 1.4). Upon division, the duplicate sets of chromosomes separate so

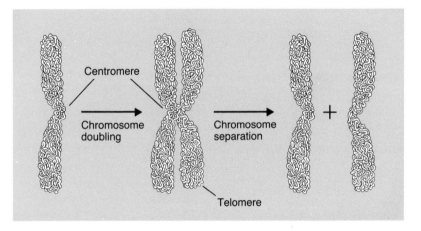

Figure 1.4   Chromosome duplication.

that each of the two daughter cells acquires the same number and kind as the parent cell had to begin with. The entire process of chromosomal division is called **mitosis**. During mitosis, the chromosomes are highly condensed and form discrete structures. Generally, the characteristic shapes and sizes of mitotic chromosomes allow homologous pairs to be identified. Actually, the chromosomes one usually sees under a microscope, and the ones normally depicted in photographs and drawings, are undergoing mitosis. Typically, the pictures show the two duplicate copies of the chromosome still linked together in mitotic chromosomes as seen in Figure 1.3.

Plate I shows photographs of lily cells undergoing mitosis; at this stage, chromosome doubling has already occurred. All of the chromosomes eventually line up in the central plane of the cell and then divide into two groups; the two groups then move to opposite ends of the cell. Two cells are formed when a membrane grows and separates the two ends of the original cell. Each of the new cells (referred to as daughter cells) has a full set of chromosome pairs.

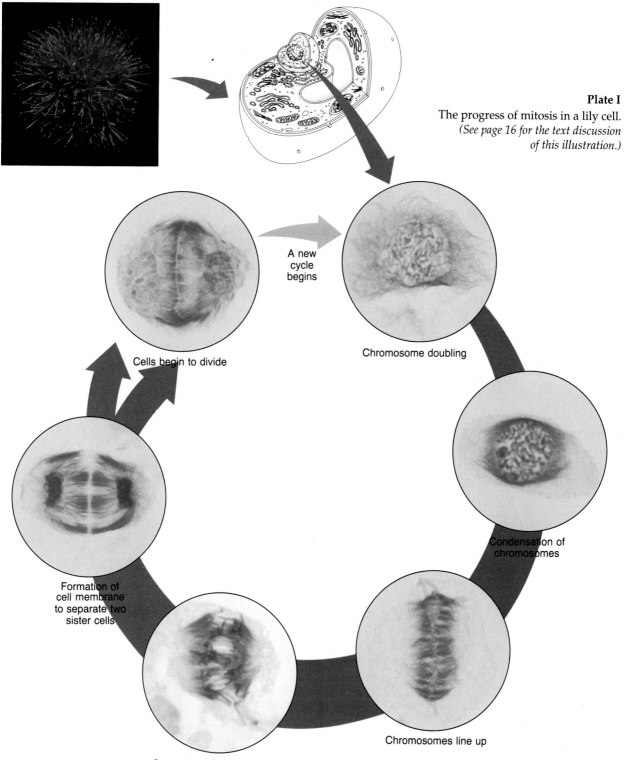

Plate I
The progress of mitosis in a lily cell.
*(See page 16 for the text discussion
of this illustration.)*

A new
cycle
begins

Cells begin to divide

Chromosome doubling

Condensation of
chromosomes

Formation of
cell membrane
to separate two
sister cells

Chromosomes line up

Separation of chromosomes

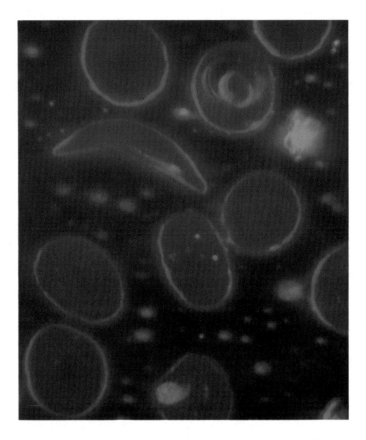

Plate II A photograph of normal and sickled red blood cells.
(See page 56 for the text discussion of this photograph.)

Plate III Examples of color variegation caused by mobile elements in maize and snapdragons. *(See page 166 for the text discussion of these photographs.)*

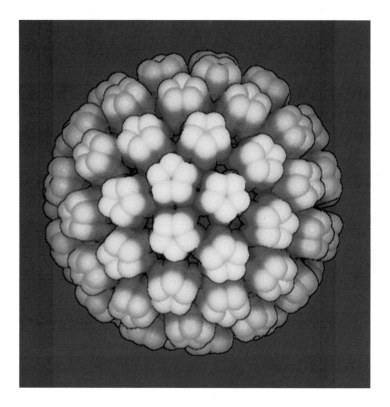

Plate IV Computer-graphics representation of a papovavirus protein coat magnified about 1 million times. (*A black-and-white version of this image is given in Figure 10.2 on page 193.*)

The events that occur from the start of one cell division to the next are termed, collectively, the **cell cycle**. Following cell separation, each daughter cell begins to make the cellular constituents characteristic of the growing cell. Following this period of active metabolic activity, chromosome doubling occurs. Once chromosome doubling is completed, the events characteristic of mitosis begin. Finally, when mitosis is completed, the two daughter cells separate and the cycle begins again. The total time for a complete cycle varies from minutes to days, depending on the cell type and growth conditions. Generally, mitosis and cell separation take up less than 10 percent of the total cycle time. During most of the cycle, the chromosomes are not readily seen because they exist in an unfolded state. **Chromatin** is the term used to refer to the unfolded state of chromosomes in the nucleus (see the cutaway of the nucleus at the center of Figure 1.1).

Productive analysis of chromosome structure depends on the choice of suitable experimental organisms. For this reason, the very bulky chromosome pairs of the salivary glands of fruit flies became a favorite experimental system early on, while systematic analysis of human and other small mammalian chromosomes had to await technical improvements made only after 1950. Even now, the small and diffuse chromosomes of such simple eukaryotes as yeasts and protozoans elude analysis by light microscopy.

The fact that eukaryotic chromosomes generally occur in homologous pairs is associated with sexual reproduction. Cells with pairs of homologous chromosomes are called **diploid**. The exceptional cells in the bodies of adult eukaryotes are the **germ cells**, the eggs and sperm. They are **haploid**—that is, they contain only one member of each homologous pair. The formation of eggs and sperm involves a reduction in the normal diploid number of chromosomes by precisely half, a process called **meiosis**. The drawing in Figure 1.5 shows meiosis in an organism like the fruit fly, with four pairs of homologous chromosomes (pairs are shown by the same shape, including the small "dot" chromosomes). The sorting of chromosomes during meiosis is random, so that either member of each homologous pair can end up in a newly formed sperm or egg. In the fruit fly and in many other organisms, including mammals, two of the male's chromosomes (termed the X and Y chromosomes) do not form a homologous pair. Thus, during meiosis, two different kinds of sperm form, one kind with an X and the other with a Y chromosome. In contrast, because the female's cells have two X chromosomes and no Y, each egg therefore contains an X chromosome. When the haploid egg and sperm fuse at fertilization, their chromosomes mingle. Consequently, fertilization reestablishes the full (diploid)

18

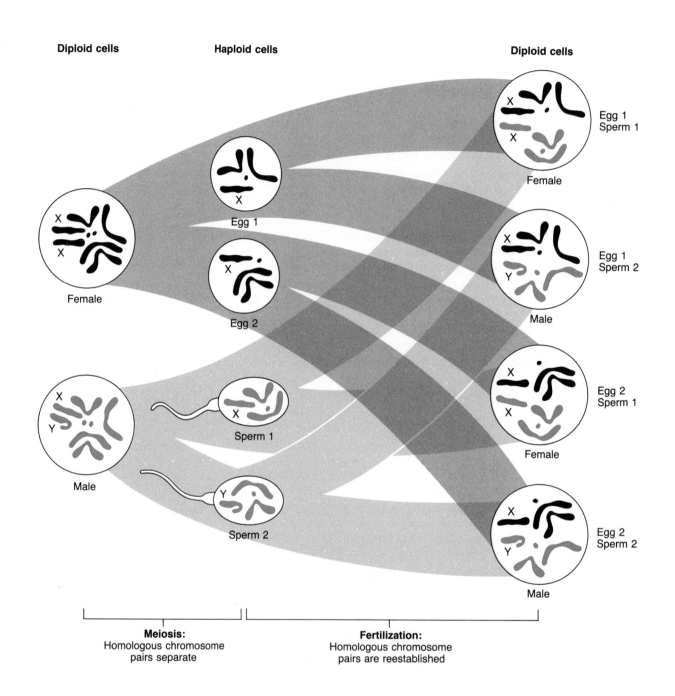

Diploid cells

Haploid cells

Diploid cells

Meiosis:
Homologous chromosome
pairs separate

Fertilization:
Homologous chromosome
pairs are reestablished

Figure 1.5   Meiosis and fertilization.

set of homologous chromosome pairs, one member of each pair being derived from the egg and one from the sperm of the respective parents. Thereafter, the fertilized egg divides by mitosis, maintaining the diploid state. The many subsequent cell divisions that can ultimately yield trillions of body cells in the mature individual are also all mitotic, except for those that yield new germ cells. The sex of the new organism that develops is determined by whether the sperm delivers an X or a Y chromosome into the egg. In some organisms (birds, for example), the situation is reversed, and it is the egg cells that are equipped with two different, sex-determining chromosomes.

The two members of a homologous chromosome pair not only look alike, they contain duplicate sets of genetic information. This fact was realized when certain organisms (sea urchins, for example) were seen to develop from eggs having only a single complement of chromosomes.

Because the chromosomes duplicate long before mitosis in the cell cycle, typical mitotic chromosomes are actually composed of two linked identical chromosomes. Figure 1.4 shows the two separate chromosomes held together at a constriction called a **centromere**. Centromeres always occur at a fixed position on each chromosome (and its homolog). However, they are located at different positions on different chromosomes. Thus, centromeres may occur in the middle, at one end, or between the middle and end of chromosomes. Very specific chromosomal functions are associated with centromeres. They are the last site of association between the two recently duplicated chromosomes before they separate during mitosis and meiosis.

All mitotic chromosomes have interesting structural features at their ends; these are called the **telomeres** (Figure 1.4). Some but not all chromosomes of a species have small constrictions at specific locations; these regions are called **nucleolar organizers**. At stages other than the mitotic phase of the cell cycle, the several nucleolar organizers are coalesced into **nucleoli** (Figure 1.1).

Special dyes and staining techniques reveal additional details of mitotic chromosomal structure. Each chromosome displays unique patterns of light and dark bands; homologous chromosomes have identical and highly reproducible patterns (Figure 1.6). By taking account of the sizes, shapes, bands, and positions of nucleolar organizers and centromeres, one can readily identify a particular chromosome and its homolog in all cells of any individual organism of the same species. Both the photographs and the diagrams in Figure 1.6 show early mitotic chromosomes from a human male, before the newly replicated chromosomes have begun to separate. The two sex chromosomes, X and Y, are obviously not alike.

20

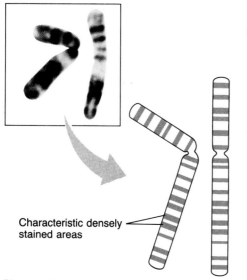

Characteristic densely
stained areas

Photograph and drawing showing location
of bands on one chromosome pair

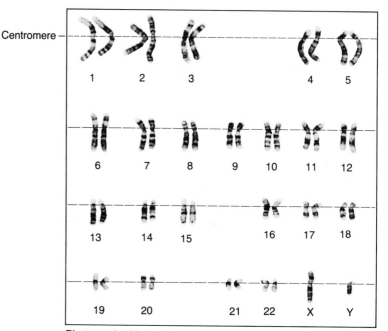

Photograph of banded human male chromosomes

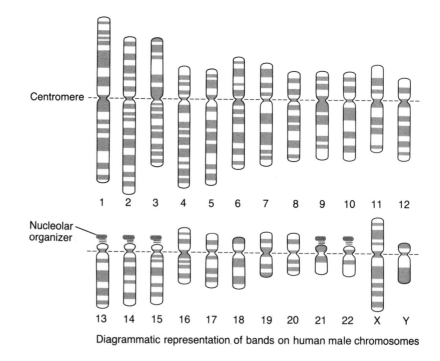

Diagrammatic representation of bands on human male chromosomes

Figure 1.6 Identification of chromosomes by their unique banding patterns.

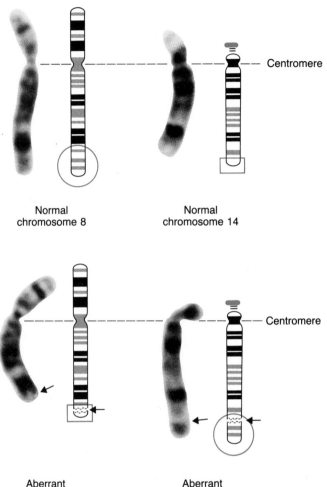

Normal
chromosome 8

Normal
chromosome 14

Centromere

Aberrant
chromosome 8

Aberrant
chromosome 14

Centromere

Figure 1.7 Rearranged chromosomes associated with a cancer.

Because the distinctive features of individual mitotic chromosomes are
the same in normal individuals of the same species, atypical numbers of
chromosomes or unusual shapes or banding patterns clearly signal an
abnormal state. Some diseases are associated with the presence of an unusual
chromosome. Figure 1.7, for example, shows an instance in which segments
of human chromosomes 8 and 14 have been exchanged (*arrows*), a rearrangement that is typically found in the cancerous cells of patients
with the disease known as Burkitt's lymphoma. In another type of chromosome abnormality, parts of chromosomes are lost. For example, the deletion of a part of human chromosome 11 is apparent from the loss of certain
bands in cells from the cancer known as Wilms' tumor. Sometimes there

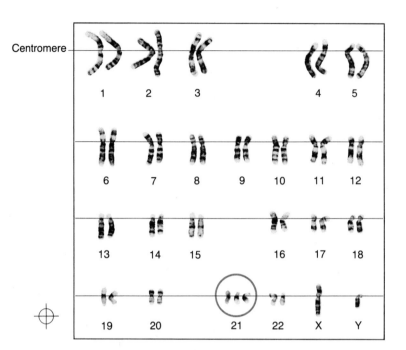

Centromere

1 2 3 4 5

6 7 8 9 10 11 12

13 14 15 16 17 18

19 20 21 22 X Y

Figure 1.8 Down syndrome is associated with an extra chromosome 21.

are extra chromosomes. For example, the human disease known as Down syndrome is associated with three copies of chromosome 21 (termed trisomy 21) rather than the usual diploid state (Figure 1.8).

Thus far, we have been describing eukaryotic chromosomes. What about prokaryotes? Recall that a prokaryotic cell (a bacterium) has no discrete nucleus (Figure 1.2). But bacteria do have chromosomes, even though they are too small and diffuse to be seen in a light microscope. Indeed, the existence of bacterial chromosomes was not known to nineteenth-century biologists, and their discovery depended on techniques developed much later. Bacteria have only a single chromosome and therefore are haploid, like the germ cells of plants and animals. Moreover, they do not undergo the complex steps of meiosis and fertilization. Instead, during the cell cycle, the chromosome duplicates, cells divide, and each of the new cells receives one complete chromosome.

Inheriting Single Traits

THE CONCEPT of the gene began with Gregor Mendel in the 1860s, although the word itself was not coined until his findings had been repeated and extended by others early in the twentieth century. The word "gene," which was introduced in 1910, referred to an abstract unit of inheritance that governs a specific trait in a particular species. The existence of genes was inferred from the statistical distribution of simple heritable traits among the offspring of known parents over several generations. Mendel, for example, studied the size, color, and texture of garden peas. Thus, the color of seeds or flowers was followed as a discernible, heritable trait without knowing what the trait actually meant in the chemistry of the organism or how the gene actually determined the color. Remarkably, the logical intellectual framework established by Mendel and those who followed him is completely consistent with modern concepts of the chemical structure of genes and how gene structure determines the attributes of an organism.

The Mendelian view of inheritance in eukaryotes was based on the existence of alternate forms of recognizable traits. For example, Mendel understood that the factor that determined the color of a pea seed could occur in different versions. In one version, it results in green seeds and in another in yellow seeds. Similarly, the seed's texture—for example, smooth or wrinkled—is also determined by alternate states of the factor determining that trait. Now that we know that genes determine these traits, we refer to the different versions of a single gene as **alleles**.

Mendel recognized that each organism contains two alleles for any single inherited trait—one from each parent. In each generation, the two allelic forms are separated (that is, they segregate independently) during formation of new germ cells. Each haploid germ cell has only one member of the original pair, one allele. Upon fertilization, a new combination of alleles is established. The two members of the pair may be the same; in this case, an individual is said to be **homozygous** for that pair of alleles. Alternatively, an individual can receive a different allele from each parent and thus be **heterozygous**. We can rephrase these concepts in a symbolic way. If we call the different alleles determining the seed color c^g and c^y (for green and yellow color, respectively), then an individual may have the following pairs: $c^g c^g$ = homozygous; $c^y c^y$ = homozygous; $c^g c^y$ = heterozygous. A germ cell will have either the c^g or c^y allele.

Mendel also discovered that pairs of alleles for different traits are independently transmitted to germ cells. For example, in a pea plant that contains the allelic pairs $c^g c^y$ for seed color and $t^w t^s$ for wrinkled (w) or smooth (s) seed texture (t), the germ cells may contain any one of the following combinations: $c^g t^w$, $c^g t^s$, $c^y t^w$, or $c^y t^s$. Thus, during the formation of germ cells, the alleles determining color, c^g and c^y, are passed on independently from the texture alleles, t^w and t^s.

Genes and Chromosomes

THE SIMILARITY between the visible behavior of the physical chromosomes during meiosis and fertilization and the purely abstract Mendelian ideas is striking. During meiosis, the homologous pairs of chromosomes separate, one member of each pair ending up in a different germ cell. Upon fertilization, a new homologous pair is established (Figure 1.5). Similarly, Mendel's theory proposed that members of allelic pairs segregate into different germ cells and new pairs of alleles are formed upon fertilization.

A now straightforward, but then creative, conceptual jump suggests that each member of an allelic pair resides on one of a pair of homologous chromosomes. Early in the twentieth century, T. H. Morgan and his colleagues at Columbia University were the first to postulate this relationship. Their ability to discern such a fundamental relationship depended on the use of the fruit fly as an experimental subject. The flies breed rapidly, with a generation time of about two weeks. Each female lays as many as 1000 eggs, so that a large number of offspring are produced from each mating. Because many different experimental matings can be carried out, genetic analysis is convenient, relatively rapid, and precise. Moreover, rare **mutations** (the sudden appearance of a new allele) can be detected because of the large number of offspring. Mendel, in contrast, had to be satisfied with one generation of peas per year and only a few offspring from each mating. Just as important to the success of Morgan and his students was the fact that fly chromosomes are readily observable in some cells with a light microscope that magnifies the image about 100 times. For the first time, correlations were made between abstract genetic analyses and actual chromosome structures.

Morgan's experiments revealed that the inheritance of an allele producing flies with white rather than the usual red eyes was always associated with inheritance of an X chromosome, never with a Y chromosome (Figure 1.9). Thus, the allele for white eyes is associated with the X chromosome. Other alleles governing other traits were also associated with the inheritance of an X chromosome. Alleles for still other traits were also frequently inherited together in linked groups but were seen to be unrelated to the X chromosome. Subsequently, it became evident that the number of groups of linked alleles equaled the number of chromosome pairs. This suggested that single chromosomes contain many genes.

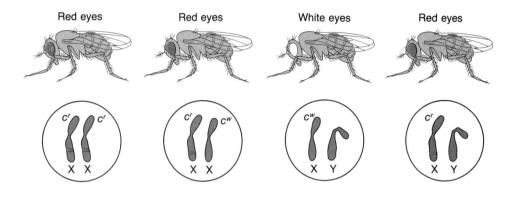

Figure 1.9 Eye color in fruit flies is determined by a gene (*c*) on the X chromosome: *c*^r and *c*^w refer to the alleles that determine the two different eye colors, red and white.

Notice that these results seem to contradict Mendel's notion of the independent assortment of alleles for different traits. Luckily for Mendel, he just happened to have studied alleles that were on different pea chromosomes; otherwise it would have been very difficult for him to discern the simple inheritance patterns that yielded such profound insight. Actually, Morgan's result refines rather than contradicts Mendel's conclusion: When alleles are associated with different chromosomes, they assort independently among the progeny (as Mendel observed); but, if alleles that determine different traits are associated with the same chromosome, they remain together in the offspring.

Almost as soon as the idea of chromosomes as groups of linked alleles (or genes) was recognized, surprising exceptions to linkage were noticed. For example, consider two unrelated hypothetical traits, *a* and *b*. Each has two different alleles—*a*¹ and *a*² and *b*¹ and *b*². When *a*¹ and *b*¹ are known to be located on the same chromosome, they are usually transmitted together to the offspring. Occasionally, however, the pairings are disrupted, and new linkages—such as *a*¹ and *b*² or *a*² and *b*¹—appear and are inherited together in subsequent generations. How could two alleles located on the same chromosome become separated in the formation of the germ cells?

Homologous chromosomes had been observed to be in contact with one another during meiosis. Figure 1.10 shows such close contacts between meiotic grasshopper chromosomes. Homologous pairs of already duplicated chromosomes are clearly discernible in several cases, and several show multiple contact points. Morgan surmised that, during such close contact, homologous chromosomes might exchange portions, thereby creating new combinations of linked alleles. This process is termed **crossing-over** or

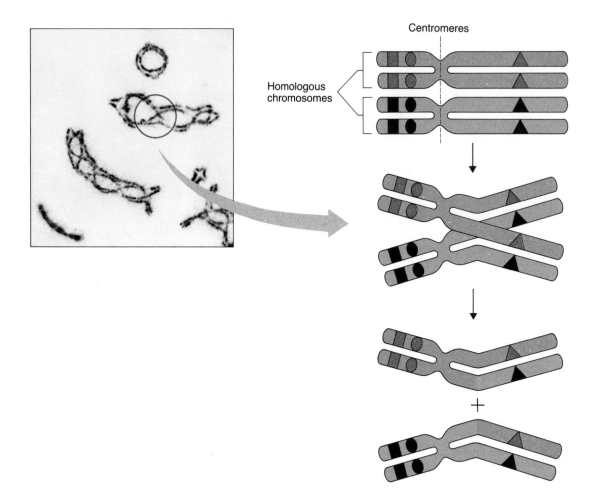

Figure 1.10 Grasshopper chromosomes in meiosis (*left*) and a diagram (*right*) showing what happens during meiotic recombination.

recombination. At the top right of Figure 1.10 are two homologous chromosomes, each a different color. The two chromosomes are already duplicated and their arms are held in proximity to one another by attachment through their centromeres. In the middle diagram, an arm of one chromosome crosses over an arm of the other. The bottom diagram shows the subsequently separated chromosomes in which one of the arms of each has recombined. In this way, new combinations of alleles for the traits indicated by squares, circles, and triangles are formed.

Without any knowledge of the underlying chemistry, geneticists began to use the phenomenon of recombination as a basic tool of genetic research.

They began to measure the frequency of recombination between linked allelic pairs. Major lasting conclusions followed, including: (1) that genes are arranged on chromosomes in a linear order, so that alleles of a particular gene usually occupy the same relative position on homologous chromosomes; (2) that recombination occurs between homologous chromosomes within the shared linkage group; and (3) that the frequency with which linked alleles of two different genes separate depends on how far apart they are on the chromosome (the farther they are apart, the more often they recombine). Using these ideas, Morgan and his colleagues were able to map the relative positions of different genes along a chromosome. By 1922, they had pinpointed the positions of several hundred genes on the four chromosomes of the fruit fly. This constituted the first successful attempt at constructing a map of the genes of any organism.

Genes and Proteins

WHAT does it mean to say that a gene determines a trait, or that an allele encodes a particular form of that trait? To trace the history of modern answers to these questions, we need to return to nineteenth-century science and its major achievements. One of the greatest of these was the recognition that cells, and thus all organisms, are complex chemical systems. Breaking open living cells and subjecting them to chemical analysis had led to a description of the cell's molecular constituents, thus allowing the chemical reactions within cells to be studied. These include, for example, reactions that permit the utilization of nutrients or the conversion of energy into usable forms for cell growth and function, or the interaction between cells to form an organized tissue.

Among the many surprises experienced by the fledgling science of biochemistry in the nineteenth century, two are especially notable. First, many biological molecules are enormous compared with the size, for example, of molecules of water or of simple sugars. The earliest of these **macromolecules** described were the **proteins**, which were shown to be made up largely of carbon, hydrogen, oxygen, and nitrogen atoms. Now we know that proteins are long chains made from 20 different small molecules—amino acids. Second, if cells are broken (by grinding, for example) and their contents are suspended in water, the solutions still promote chemical reactions initially thought to occur only in intact living organisms. Early examples included the conversion of starch to sugar by extracts of germinating barley, or the breakup of coagulated egg white by water extracts from an animal's stomach lining.

The analogy between these reactions and the then recently discovered phenomenon of **catalysis** of chemical reactions was promptly discerned by nineteenth-century chemists. Catalysis had been defined as a process in which a chemical reaction is speeded up, sometimes many thousandfold, by an additional chemical agent, the catalyst, which is not itself altered in the reaction. Cell extracts were thus inferred to contain biological catalysts, which were termed **enzymes**. At about the same time, it became clear that enzymes are proteins, a conclusion that was amply confirmed in 1935 when J. H. Northrop, working at the Rockefeller Institute for Medical Research in New York, demonstrated that a pure protein prepared from the stomach lining of cattle was identical with the digestive enzyme pepsin.

The first clue as to how genes determine traits came from the study of enzymes. The clue was discovered even before the invention of the word *gene*. In the first decade of the twentieth century, the English physician Archibald Garrod noted that the inheritance of certain human traits and diseases followed Mendelian rules. Even without experimental proof, the proposal was attractive. It fostered a sense of the unity of nature by relating the then contemporary genetic studies on flies and plants to human biology. Garrod surmised that these inherited characteristics stemmed from the deficiency or absence of particular enzymes required for normal cell chemistry. He then suggested that the determinants of heredity controlled the production of enzymes. Thus, the ability of an organism to produce enzymes was attributed to its genes. More specifically, Garrod supposed that the relative efficiency with which an enzyme worked was dependent on the status of the gene that governed the enzyme's production.

To pursue Garrod's idea successfully required new experimental methods. The use of single-celled microorganisms as experimental subjects, beginning in the late 1930s, provided such a new approach. Initially, simple fungi were the focus of attention. These organisms grow readily in liquid containing simple but defined nutrients. They also reproduce quickly enough to yield many generations of descendants within days. With such systems, researchers were not confined to studying directly observable traits, such as color, but could study inherited nutritional traits as well. An example of such a trait is whether the cultured cells need a particular nutrient for growth or can dispense with it entirely. Sufficient genetic and biochemical data were obtained and correlated by the middle 1940s to establish that the presence or absence of an enzyme was heritable and governed by the function of a single gene. By the early 1940s, George Beadle and Edward

Tatum, both professors at Stanford University in California, had generalized the relation between enzyme and gene as "one gene—one enzyme." Because enzymes were known to be proteins, and because some genes specify proteins that are not enzymes (for example, some function in muscle contraction, others constitute part of the skeleton, and still others, such as insulin, function as hormones), this notion evolved to "one gene—one protein." But even this representation of the gene–protein relationship is inadequate, as will be evident when the details of protein structure are described in Chapter 2. Moreover, a few genes dictate the formation of molecules that are not proteins at all (Chapter 3).

The two components in this informational relationship are a gene and the protein it specifies. It is proteins that give an organism and its cells their characteristic shapes and chemical, physical, and behavioral capabilities. Such characteristics are frequently the result of an interplay between an organism's genes and the environment in which that organism exists.

By 1950, an even more attractive system had been discovered for examining the relationship between genes and biological functions. A common intestinal bacterium, *Escherichia coli* (familiarly known simply as *E. coli*), has simple nutritional requirements. A single cell duplicates its single chromosome and divides into two cells every 20 to 60 minutes, depending upon the growth conditions. Thus, one can obtain large numbers of cells (a billion per milliliter of culture fluid, or 30 billion per ounce) in less than 24 hours. A large number of readily measurable *E. coli* nutritional requirements were found to be genetically controlled. Furthermore, readily obtainable nutritional mutants provide easily detectable alleles of *E. coli* genes. For instance, an *E. coli* cell that is unable to synthesize an essential cellular component will be unable to produce a visible colony of cells on an otherwise complete growth medium lacking that essential component. However, the same cell can grow if its medium is supplemented with the essential ingredient. This fact allowed researchers to establish correlations between particular genes and specific cellular functions. Because the genetic methods used with fruit flies were not applicable to bacteria, special tricks were needed to construct a genetic map of the single chromosome of *E. coli*.

The mapping techniques devised by Morgan depended on recombination between the pairs of homologous chromosomes typical of sexually reproducing diploid organisms. *E. coli* and other prokaryotic cells reproduce by doubling their single chromosome and then dividing in two, with one copy of the chromosome going into each cell. Thus, prokaryotes are always haploid and, therefore, although different alleles of prokaryotic genes exist, they occur in different cells. For example, a gene that permits the bacteria to use the simple sugar glucose as a nutrient may occur as a functional allele or as a nonfunctional mutant allele. Judging whether or not a bacterial cell can grow on glucose allows one to determine which of the two alleles that cell contains. Thus, it became possible to exploit fully the ability of bacteria to produce many offspring rapidly and asexually and to use the many mutant forms of *E. coli* for genetic analysis. A useful trick was found in the observation that, under some circumstances, prokaryotes can become partially diploid (for more details, see Chapter 4). Here, the important point is that genetic analysis of *E. coli* clearly showed that, even in such simple organisms, the order of individual genes on the single chromosome could be determined.

Genes Are DNA

TODAY'S genetic chemistry has its origin in 1869, when Johann Friedrich Miescher, a German chemist working in Tübingen, discovered that a material extracted from human pus cells and cell nuclei was chemically different from proteins both because it contained phosphorus and because it resisted destruction by such enzymes as pepsin. He called the material "nuclein"; now we refer to it as **deoxyribonucleic acid** or **DNA**. During the next 85 years, new methods were developed to purify DNA and to establish the chemical nature of its constituents and the nature of the bonds between them. These investigations identified the basic structural units of DNA: four distinct, small molecules called (as a group) **nucleotides**. Meanwhile, studies of the physical properties of DNA established that it is a polymer—that is, it consists of very long, chainlike molecules whose individual units are nucleotides. Moreover, it was found that most of the DNA in cells is contained in chromosomes. In 1953, the existing chemical and physical information about the composition and behavior of DNA was assimilated by James Watson and Francis Crick, two young scientists working at Cambridge University in England, in their now classical double-helical structure. Figure 1.11 shows that a DNA double helix consists of two separate, long, chainlike molecules that wind together around a common axis. The two chains are held together by weak chemical interactions between the nucleotide units on each chain. Chapter 2 explains more about the structure of the double helix. The linear order of the four kinds of units, indicated in Figure 1.11 by the four different shapes, constitutes the genetic information. How that genetic information is used to produce proteins is explained in Chapter 3.

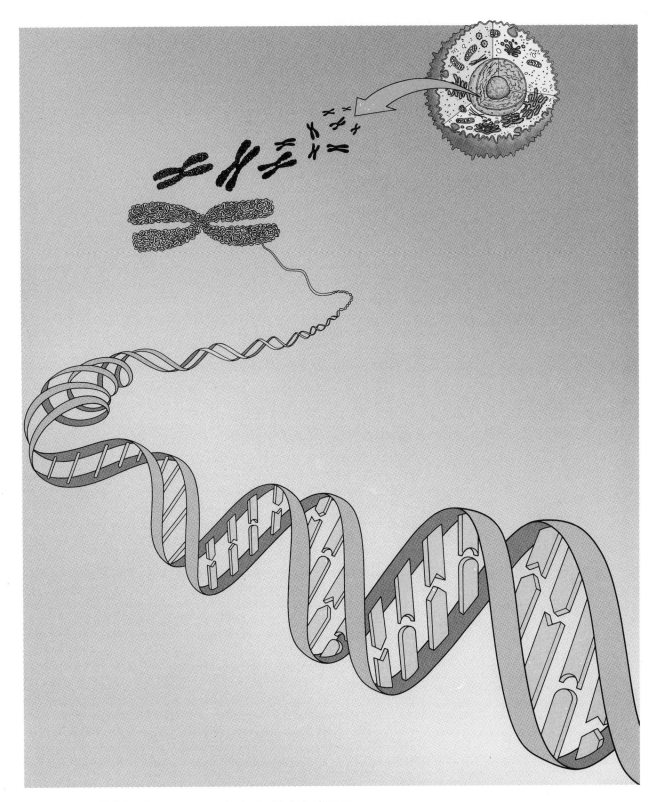

Figure 1.11 Cellular chromosomes contain double-helical DNA.

The significance and impact of the double helix were magnified by the growing acceptance of the discoveries by Oswald Avery and his colleagues (at the Rockefeller Institute in New York) and by Alfred Hershey and Margaret Chase (at Washington University in St. Louis, Missouri) that DNA alone was the carrier of genetic information. The central role in heredity previously attributed to chromosomes could now be confidently assigned to chromosomal DNA. The many proteins that are also contained in chromosomes are important, but they do not contain hereditary information.

DNA is not the only nucleic acid in cells. Closely related nucleic acids, molecules of **ribonucleic acid (RNA)**, occur both in the nucleus and in the cytoplasm of cells. As we will see, RNA plays many roles critical to the way in which the cell utilizes the information in the DNA structure to make proteins.

Genes Are Molecules

THE DESCRIPTION of the structure of DNA and the recognition of the central role of DNA in heredity represent the crowning achievements in the transition of genetics from a statistical and phenomenological science to one with an overriding chemical and molecular perspective. All at once, heredity could be correlated with molecular structure. Almost a century after their independent beginnings, the three separate scientific fields—the study of chromosome behavior and structure, abstract genetic analysis, and biochemistry—were unified. The immediate, exuberant response to the proposal of the double helix reflected a sense of its correctness. Not only did the model fit the chemical and physical data, but the structure was perfectly suited to the functions required of the genetic material. The linear sequence of a DNA chain's four different nucleotides could represent what we have referred to as genetic information. Thus, just as a book encodes information in the linear sequence of the alphabet's 26 letters, the linear order of the four nucleotides can provide a genetic message for producing proteins. The double-helical nature of DNA also enables one to see how DNA can direct its own replication prior to cell division (see Chapter 2), so that each new cell acquires a full set of genetic instructions. DNA had all the attributes required for it to function as the substance of genes.

The biological principles established over a hundred years of research can now be restated in molecular terms. Not only do homologous chromosomes look alike, they contain DNA molecules with very similar or identical sequences of the four different nucleotides. A gene is no longer an abstract entity: It is a sequence of nucleotides, a segment of a DNA chain. Alleles differ from one another in nucleotide sequence; a change in a gene's nucleotide sequence may alter its function and cause an altered trait. Such changes in the DNA sequence, especially if they cause noticeable changes in an organism's characteristics, are called **mutations**. New mutations generate new alleles. Recombination between homologous chromosomes during meiosis is, at the molecular level, the breaking of two double-helical DNA molecules and the rejoining of the pieces of one with the pieces of the other so as to create new DNA combinations. Chromosome doubling in mitotic cell cycles and in meiosis reflects the duplication (or **replication**) of DNA double helices. A relatively new term, **genome**, was introduced to describe the totality of the DNA present in the chromosomes typical of each species. The genome carries almost all of the information necessary to define an organism's properties.

Understanding the structure of DNA illuminated every aspect of biology and provided a foundation upon which the protean observations of preceding centuries could settle. It defined a fundamental unity for interpreting the enormous diversity of life.

CHAPTER 2

Molecules Convey Information

B EGINNING in the early 1950s, the long-standing questions concerning the mechanisms of heredity were reformulated in chemical terms. How do DNA molecules replicate and recombine? Why are mutations maintained in succeeding generations? How does the genetic information contained in DNA determine biological structure and the chemical processes within cells? Is the flow of information from DNA regulated during cell growth, development, and other physiological states? And how are these processes altered in disease? These questions have helped form the agenda for research in molecular genetics during the last 40 years. The explosive progress during the first 20 of those years occurred primarily in studies with prokaryotic systems and provided sound answers to the basic questions.

Three kinds of molecules are central to genetic processes: DNA, RNA, and protein. The informational relationships between these genetic molecules are summarized in Figure 2.1. Genetic continuity from one generation to the next requires that DNA be reproduced for delivery into new cells

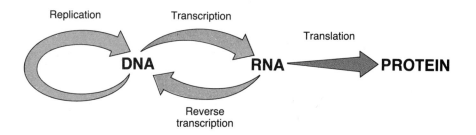

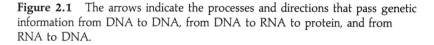

Figure 2.1 The arrows indicate the processes and directions that pass genetic information from DNA to DNA, from DNA to RNA to protein, and from RNA to DNA.

with each cycle of cell division. **DNA replication** is the process by which the parental DNA molecules are duplicated before being passed on to each of the offspring. This replication process must occur with high fidelity to avoid passing on misinformation. Moreover, damage (such as that caused by ultraviolet light) and inadvertent errors (such as the introduction of a wrong nucleotide into a DNA sequence) between and during replication cycles need to be rectified rather than passed on to the new cells. Thus, DNA is the subject of a variety of transactions—replication, repair, recombination, and even rearrangement. These are the key processes by which organisms maintain and diversify their genomes.

DNA carries genetic information in the order of the four nucleotides in its chains. This is analogous to the way that written information is represented by the order of letters on a page. Just as a written sentence contains one idea, a gene represented by a portion of a DNA molecule contains one unit of genetic information. Cells must decipher the information for the corresponding trait to be revealed. The process of deciphering a gene's information involves several steps, which are collectively called **gene expression**. In the first step, the order of different nucleotides in a gene is copied into a related nucleic acid molecule called RNA, a process called **transcription**. In the second step, the RNA directs the production of another kind of molecule, a protein; in this process, called **translation**, the order of nucleotides in the RNA determines the nature of the protein that is made. Because the order of nucleotides in each gene (and in the RNA transcribed from it) is different, each gene directs the formation of a different protein. Often, this idea is abbreviated by saying that a gene encodes a protein. Each cell's and organism's characteristics are, therefore, ultimately determined by the number and variety of proteins decoded from its DNA.

As we shall see in Chapter 3, DNA is transcribed into several kinds of RNA, only one of which is translated into proteins. Other kinds of RNA participate in the chemical processes that cells use to manufacture proteins.

Fundamentally, genetic information flows in one direction, from DNA to RNA to protein. In certain specialized cases, information can also pass from RNA back into DNA by a process called **reverse transcription**. So far as is known, information stored in proteins is not used to assemble corresponding nucleic acids; in other words, reverse translation is unknown. Nevertheless, as we shall see, proteins are critical participants in the processes that transfer information between nucleic acids and subsequently to proteins as well.

A central feature of information transfer between nucleic acids, whether it be replication, transcription, or reverse transcription, is the function of the nucleic acid as a **template** to direct the assembly of new chains with related nucleotide sequences. The basic notion is that the order of nucleotides A, T, G, and C on an existing chain—the template—determines the order of nucleotides on the newly made chain.

Some understanding of the molecular structures of DNA, RNA, and proteins is necessary to make sense of their informational relationships. This is because the replication of DNA and the decoding of genetic information depend on chemical reactions, as do all the other cellular processes that together make a cell alive. This chapter describes the key structural features of the genetic macromolecules. This will provide a basis for the later description of how these macromolecules function in replication and gene expression. The fundamental structural features of DNA, RNA, and protein are the same in all organisms, prokaryotes and eukaryotes alike. This is a remarkable demonstration of the unity of the living world and also a welcome simplification in thinking about biology.

Atoms and Molecules

BEFORE descriptions of complex biological macromolecules are given, it may help some readers to review a few basic chemical concepts. More knowledgeable readers can skip this brief section.

All substances in the universe are composed of a limited number (about 100) of different kinds of atoms. Only a few of these atoms are major components of living organisms on the planet Earth. They include (with their chemical symbols given in parenthesis) carbon (C), oxygen (O), nitrogen (N), phosphorus (P), hydrogen (H), and sulfur (S). Other atoms essential to life, such as magnesium (Mg), calcium (Ca), iron (Fe), sodium (Na), potassium (K), and chlorine (Cl), are also abundant. Still other atoms, while essential, occur only in trace amounts in cells; these include zinc (Zn), cobalt (Co), and manganese (Mn).

When atoms combine, they form molecules. For example, gaseous oxygen consists of molecules containing two oxygen atoms each. Each water molecule contains two hydrogen atoms and one oxygen atom. Gaseous oxygen and water are small molecules. But very large biological molecules are also composed of atoms. A DNA molecule contains atoms of carbon, hydrogen, oxygen, nitrogen, and phosphorus. Depending on how long the DNA chain is, the total number of atoms in the molecule may be millions or even billions.

The structure of an individual atom can be imagined as being somewhat like the solar system (Figure 2.2). At the center, in the sun's position, is the atom's nucleus. Electrons are the other component of atoms. They orbit the atomic nucleus in ways resembling planets orbiting the sun, but they trace out spheres surrounding the nucleus. When atoms are joined to form molecules, the electrons belonging originally to individual atoms now orbit together around the nuclei in complex paths. This holds the atoms together in what is called a **chemical bond**, which is generally represented on paper by a straight line between atoms. In gaseous hydrogen, for example, the two hydrogen atoms share electrons, forming a chemical bond. In larger molecules, only neighboring atoms form bonds with one another. For example, alcohol has two carbon, six hydrogen, and one oxygen atom (Figure 2.3). The sugar present in DNA, deoxyribose, contains a more complex assortment of carbon, hydrogen, and oxygen atoms.

Most biological molecules are quite stable, although some are more stable than others. Energy is required to make the chemical bonds that hold molecules together. Energy is also required to break bonds, more being required to break strong bonds than to break weak ones. The energy can be supplied in various forms, including heat.

DNA

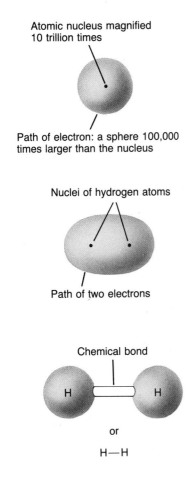

Atomic nucleus magnified 10 trillion times

Path of electron: a sphere 100,000 times larger than the nucleus

Nuclei of hydrogen atoms

Path of two electrons

Chemical bond

H — H

or

H—H

a

b

c

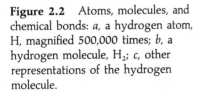

Figure 2.2 Atoms, molecules, and chemical bonds: *a*, a hydrogen atom, H, magnified 500,000 times; *b*, a hydrogen molecule, H₂; *c*, other representations of the hydrogen molecule.

IMAGINE an ordinary chain made of links. Now hang a pendant from each link. This is the form of a DNA chain, also called a single DNA strand. Each link with its pendant is a nucleotide unit. The nucleotide units are found as individual molecules in cells and become linked together to form DNA strands.

Four different kinds of nucleotide molecules are found in DNA. The four are very similar in chemical structure but different enough to be distinguished by the processes that use them (Figure 2.4). The distinctive parts of the four nucleotides are the rings called bases, formed by chemical bonds between carbon and nitrogen atoms; other atoms are joined directly to the rings. The four different rings are called adenine, thymine, guanine, and cytosine, and the corresponding nucleotide units are referred to by the letters A, T, G, and C. In a nucleotide unit, the distinctive ring is joined to another group of atoms that is identical in each kind of nucleotide. This common portion includes a sugar called deoxyribose (also a ring) attached

Name:	"Ball and stick" representation	"Space filling" representation

Oxygen

Water

Alcohol

Deoxyribose

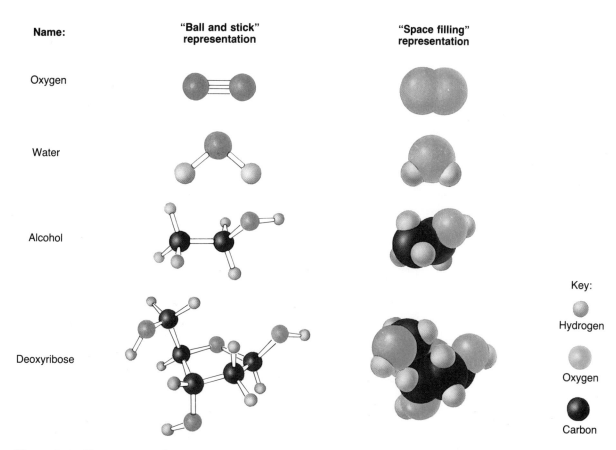

Key:

Hydrogen

Oxygen

Carbon

Figure 2.3 Various ways of representing molecules.

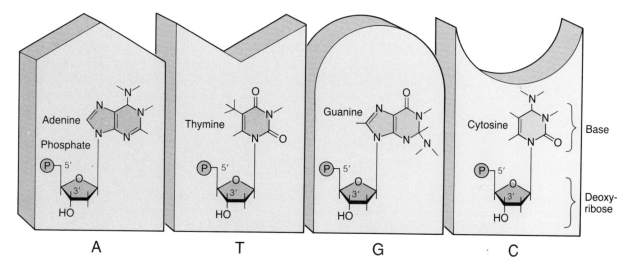

Figure 2.4 Nucleotides, the basic structural units in DNA. Note that a carbon atom is present (though not shown) at each point where two or more chemical bonds (straight lines) intersect. Bonds with no designated atom at one end are understood to be joined to a hydrogen atom.

to a phosphate (P) group. The distinctive shapes shown for the four nucleotides in Figure 2.4 will be used in many of the illustrations of nucleic acids.

Long DNA strands result from forming chemical bonds between the phosphate on each deoxyribose and the deoxyribose of a neighboring nucleotide (Figure 2.5). Thus, the nucleotide units are held together through their phosphates. Returning to the original analogy, the deoxyribose phosphate elements form the chain, and the bases are the pendants.

In a DNA strand, each deoxyribose is connected to two different phosphates. The two deoxyribose phosphate links are to different carbon atoms (labeled 3′ and 5′ on Figure 2.5) on the deoxyribose and are not equivalent; this can be seen by careful inspection of Figure 2.5. In the left-hand strand, each nucleotide is linked to the phosphate below through a deoxyribose carbon designated 3′ (and pronounced "three-prime"), and to the phosphate above through a deoxyribose carbon designated 5′. The opposite is true on the right-hand strand. The details are not relevant here, but it is important to keep in mind that DNA strands have directionality as a consequence of these different linkages. By convention, reading up the left-hand strand or down the right-hand strand is referred to as reading in the 3′ to 5′ direction, and reading down the left or up the right is 5′ to 3′. This directionality of DNA strands is confusing, but it will soon become clear why it is interesting.

Cellular DNA consists of a combination of two single DNA strands. The two chains are held together by weak chemical bonds called **hydrogen bonds** (shown as dots on Figure 2.5), which hold nucleotide bases on one strand to nucleotide bases on the other. Only certain pairs of nucleotides can form such bonds: T always pairs with A, and C always pairs with G. A consequence of these virtually invariant base pairs is that the sequence of bases on one strand uniquely defines the sequence of bases on the other strand to which it binds. Such paired DNA strands are said to be **complementary** to one another. The complementary strands go in opposite directions. One chain goes from 5′ to 3′ and the other 3′ to 5′. Note too that complementary strands generally have different nucleotide sequences—for example,

$$\overrightarrow{\text{-ACTGA-}}$$
$$\overleftarrow{\text{-TGACT-}}$$

in which the arrows indicate the opposite directions of the two strands.

The two complementary DNA strands wind around a common axis to form the famous double helix. Tracing the outside of the helix are the

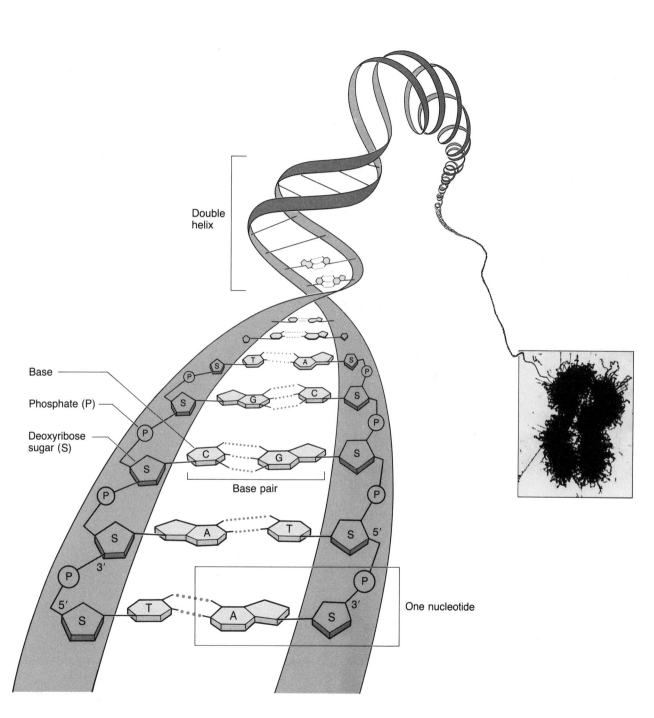

Base

Phosphate (P)

Deoxyribose
sugar (S)

Double
helix

Base pair

One nucleotide

Figure 2.5 The arrangement and association of nucleotides in the DNA double helix.

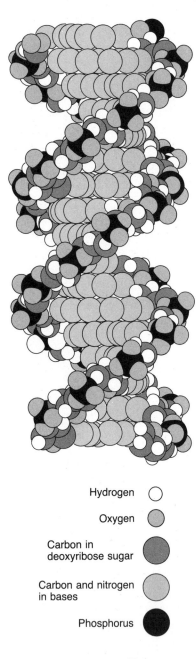

Hydrogen ○

Oxygen ◐

Carbon in
deoxyribose sugar ●

Carbon and nitrogen
in bases ○

Phosphorus ●

Figure 2.6 A space-filling representation of the atoms in a DNA double helix.

long repeating stretches of deoxyribose phosphates (this is seen clearly in the representation in Figure 2.6). Pointing inward are the bases, with the weak hydrogen bonds between paired bases on the two strands holding the double helix together. Double-helical DNA molecules are long, flexible, threadlike structures, with a nearly constant diameter, regardless of the order of the four nucleotides. Figure 2.7 shows DNA threads that have been extruded from a bacterial virus (T2 phage), a bacterial cell (*E. coli*), and a human chromosome. Each of the threads shown in the pictures in Figure 2.7 consists of a DNA double helix of the form shown in Figures 2.5 and 2.6. DNA molecules differ in length according to the number of base pairs they contain. A molecule that is one thousand base pairs long has a length of 0.00034 millimeter. Thus, for example, all the DNA in a single human cell—about 6 billion base pairs—would be more than 2000 millimeters long if it were stretched out. In chromatin and chromosomes, however, the double helices are neatly wound and folded, so that all the human DNA fits readily into a cell's nucleus as small as 0.005 millimeter in diameter.

Each eukaryotic chromosome contains a single long, linear DNA double helix. However, some other DNA molecules are circular and therefore have no ends. For example, bacterial chromosomes are often circular and relatively small. The common bacterium *E. coli* has a single circular chromosome (the DNA of which is shown in Figure 2.7) containing about 4 million base pairs, and it is thus about 1.4 millimeters in circumference. It is folded to fit into a cell no bigger than 0.001 millimeter across.

Figure 2.8 shows photographs of linear and circular DNA molecules taken with an electron microscope. Below the photographs are drawings of DNA illustrating different representations. Usually, DNA chains are represented horizontally on the page (rather than vertically, as in Figures 2.5 and 2.6). Often only the sequence of nucleotides is illustrated, as in "AAGCTG," and so on. Direction is shown by indicating the 5' to 3' orientation: 5'-AAGCTG-3'. Note that a double-helical region and its complementary base sequence can be written as follows:

$$5'\text{-AAGCTG-}3'$$
$$3'\text{-TTCGAC-}5'$$

The four nucleotides, A, G, T, and C, do not occur in equal amounts in most DNAs. Because of base pairing, however, there are certain invariant quantitative relationships, irrespective of the source of the DNA. The amounts of A and T are always equal, as are the amounts of G and C because an A always pairs with a T and a G always pairs with a C. Actually, the fact that the amounts of A and T or G and C are equal was discovered

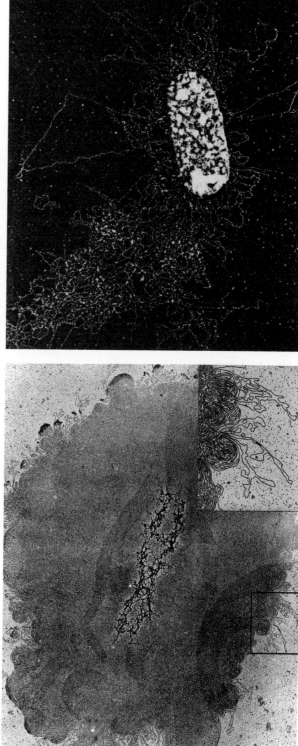

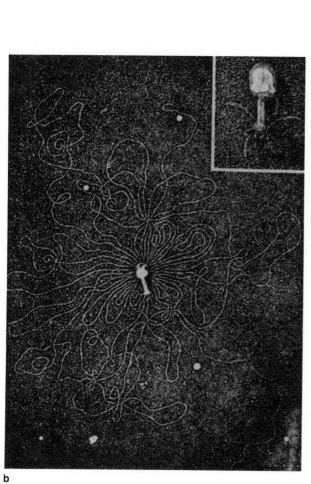

Figure 2.7 Electron microscope images showing the threadlike structure of DNA: *a, E. coli* DNA; *b,* T2 phage DNA; *c,* human DNA.

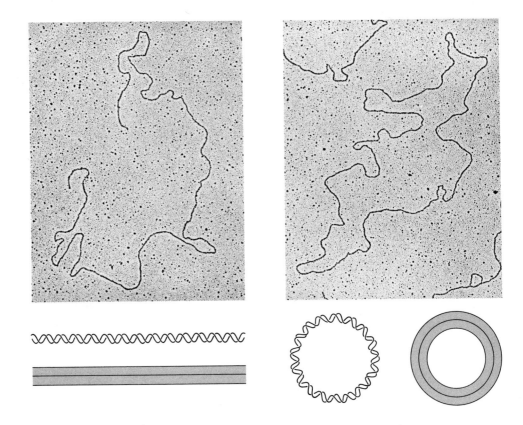

Figure 2.8 Linear (*left*) and circular (*right*) double-helical DNA.

before the double helix was proposed, and it was a critical clue to solving the DNA structure.

Because only relatively weak chemical bonds hold the two chains of the double helix together, the two strands can be unwound and separated with an input of rather a small amount of energy. Heating a solution of DNA in water close to the boiling point rapidly breaks the hydrogen bonds and unwinds the double helix without breaking the stronger bonds that hold the nucleotides in the single strand together (Figure 2.9). Such a disrupted DNA is said to be **denatured**. If the temperature is lowered, the process is reversed. Under the right conditions, the two single strands will realign properly and reform the original base pairs and the double helix, a process called **annealing**.

44

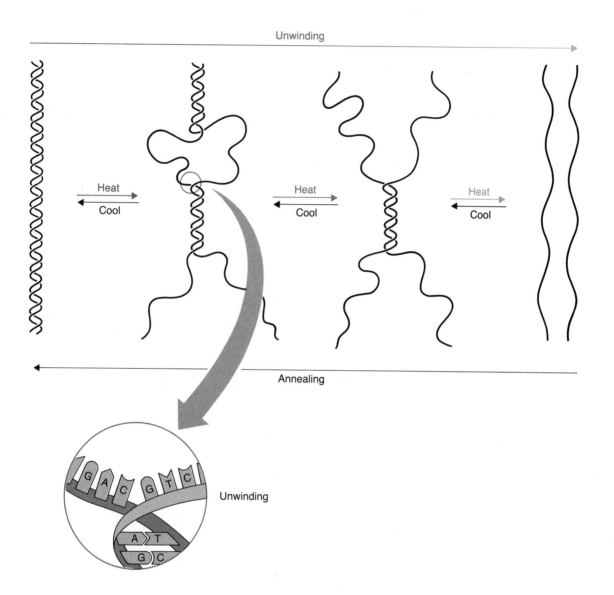

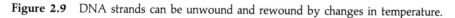

Figure 2.9 DNA strands can be unwound and rewound by changes in temperature.

Unwinding and annealing of DNA strands is an artificial, laboratory reenactment of a process that is common and critical to the biological functions of DNA. It is also a critical component of the laboratory procedures used in studying and manipulating DNA. We will come back to this process repeatedly, so it is well to keep in mind that the fundamental requirement for annealing two DNA strands is that their nucleotide sequences be complementary. The order of the nucleotides—A, T, G, and C—on one strand

45

must be matched by a corresponding order of the complementary nucleotides—T, A, C, and G—on the other, with the two strands going in opposite directions.

Studies of simple prokaryotic genomes revealed that DNA molecules play two different roles in their own replication. First, the sequence of bases on each chain provides a template from which a new strand is copied. Second, genes within the DNA encode enzymes and other proteins that are required for new DNA synthesis. During replication, the two strands unwind at a particular site in the DNA double helix (Figure 2.10). Each strand then acts as a template for the formation of a new complementary strand. Thus, each of the two newly formed double helices inherits one of its strands from the parental DNA helix, the other strand being newly synthesized. Although simple in its logic, DNA replication is actually a complex process requiring many proteins, among which the enzymes called **DNA polymerases** are central. (The suffix *-ase* is commonly used to denote an enzyme.) The role of DNA polymerases is to assemble the new DNA strands from individual nucleotide units. All DNA polymerases elongate strands one nucleotide at a time. The nucleotide to be added at each step must be complementary to the next available base in the template strand so that it can form the appropriate hydrogen bonds to hold the strands together. The high fidelity of the replication process assures a nearly error-free transmission of genetic information from one generation to the next.

DNA replication may start at multiple sites in a very long double helix. Moreover, for reasons related to the details of the process, only short segments of the new strands are produced. Therefore, replication requires the joining of many short DNA strands to form the very long chromosomal DNA. Another enzyme, **DNA ligase**, catalyzes this joining, forming standard links between DNA fragments. Not surprisingly, given their very important functions, both DNA polymerases and DNA ligases are found in all cells.

DNA molecules can undergo a variety of rearrangements. Pieces of DNA can move from one place to another on chromosomes. Pieces of DNA can also be lost, and sometimes a segment of DNA is repeated. Oftentimes such rearrangements are disadvantageous. Occasionally, however, they create new combinations of genes that allow for evolutionary experimentation. Among the enzymes and proteins encoded in all genomes are those that participate in these rearrangements. One of these processes, recombination, produces the exchanges between DNA segments on homologous chromosomes that occur during meiosis (described on page 25 and in Figure 1.10). In molecular terms, recombination results from the

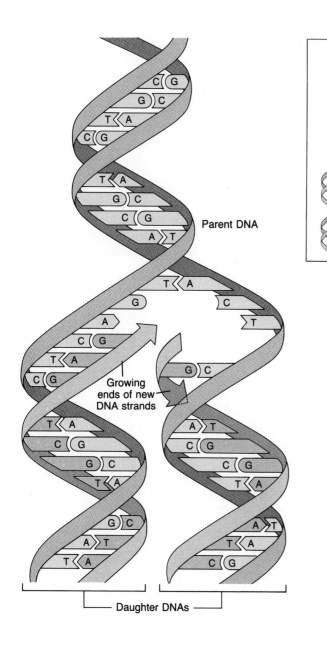

Parent DNA

Growing
ends of new
DNA strands

Daughter DNAs

Figure 2.10 A replicating DNA molecule.

breakage and rejoining of DNA strands (Figure 2.11). Recall that the nucleotide sequences in the double-helical DNAs in each of a pair of homologous chromosomes are very similar, usually almost identical; the sequences will differ slightly if two different alleles (lowercase and capital letters in Figure 2.11) of a gene happen to occur. The breaks and joinings involved in recombination generally occur at identical nucleotide sequences (frequently referred to as homologous sequences) within the pair of double helices.

DNA also encodes enzymes and other proteins that foster the repair of damaged DNA. Exposure of cells to ultraviolet light or x rays or to a variety of chemical agents produces a wide assortment of molecular lesions in the bases or in the deoxyribose phosphate portions of the DNA helix. There are enzymes and related proteins that repair such damage and errors made during replication. Thus, organisms have evolved complex mechanisms for maintaining the integrity of their genomes.

RNA

RNA chains are very similar to single DNA strands (Figure 2.12). There are two major differences. First, the sugars in RNA are ribose, rather than deoxyribose (ribose contains an OH instead of the hydrogen in deoxyribose). Second, the base uracil (nucleotide U) replaces thymine (nucleotide T). RNA chains range in size from less than 100 to tens of thousands of nucleotides. The links that hold the nucleotides together are the same as in DNA: a phosphate links the 5' position of one ribose ring to the 3' position on the ribose of the neighboring nucleotide.

RNA forms the bulk of the nucleic acid in all cells, being five to ten times more abundant than DNA. As summarized in Table 2.1, there are several different kinds of RNA, most of which occur in all cells and play particular roles in translation—that is, the synthesis of proteins (see Chapter 3). By far the most abundant RNAs are the several types called **ribosomal RNA (rRNA).** They occur in all cells, although their structures vary somewhat among different species of organisms. In any given species, each of the ribosomal RNAs has a fixed nucleotide sequence, and thus the thousands of copies of each in a cell are identical. On the other hand, the molecules of **messenger RNA (mRNA)** in a cell constitute an enormously complex mixture of molecules with different sequences of the A, U, G, and

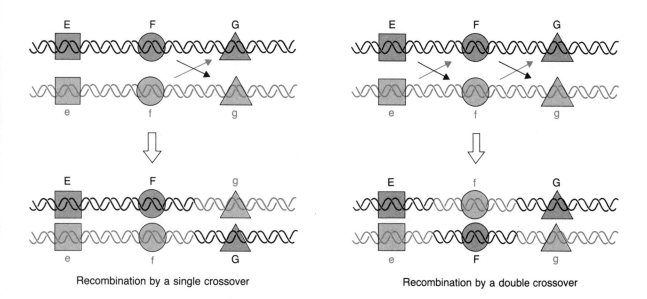

Recombination by a single crossover   Recombination by a double crossover

Figure 2.11 Recombination: the exchange of DNA segments between homologous DNA molecules.

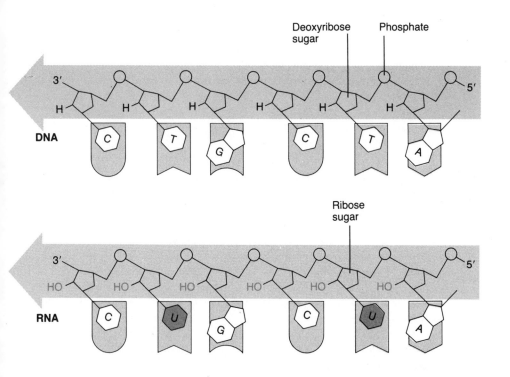

Figure 2.12 Similarities and differences between DNA and RNA.

Table 2.1 *Some Important Cellular RNAs*

	Approximate number of different kinds in cells	Approximate length (in nucleotides)
Transfer RNA	40−60	75−90
5S ribosomal RNA	1−2	120
5.8S ribosomal RNA*	1	155
Small ribosomal RNA	1	1600−1900
Large ribosomal RNA	1	3200−5000
Messenger RNA	thousands	vary
Heterogeneous nuclear RNA*	thousands	vary
Small cytoplasmic RNA	tens	90−330
Small nuclear RNA*	tens	58−220

* Occurs only in eukaryotic cells

C nucleotides. The explanation for these differences will emerge when the functions of the different RNA types are described in Chapter 3.

Unlike DNA, most cellular RNA occurs as single strands. However, within many RNA molecules there are short sequences of nucleotides that are complementary to other sequences in the same strand. Such pairs of complementary sequences can form hydrogen bonds when they come in contact with each other. [For example, a 5'-UAUUC-3' sequence can pair with a 3'-AUAAG-5' sequence somewhere else on the strand, if the strand loops back on itself.] Frequently, such intramolecular folding is critical to RNA function. The best understood molecules in which such structures are important are known as **transfer RNA (tRNA)**. A diagram of the folded structure of a tRNA molecule is shown in Figure 2.13.

The complementary base pairs in a folded single strand of RNA can be broken by heat in much the same way as those in a DNA double helix can be. In principle, RNA chains can even form a long double helix, but such molecules are rare in cells because usually no complementary strand exists. Unlike DNA replication, RNA synthesis produces only single strands (Chapter 3).

If an RNA strand and a single DNA strand have complementary base sequences, they can form an RNA–DNA double helix when conditions are right. Here, the uracils in RNA pair with the adenines in DNA, while the thymines in DNA pair with the adenines in RNA. Such hybrid pairing is an essential tool in the laboratory manipulation of nucleic acids. For example,

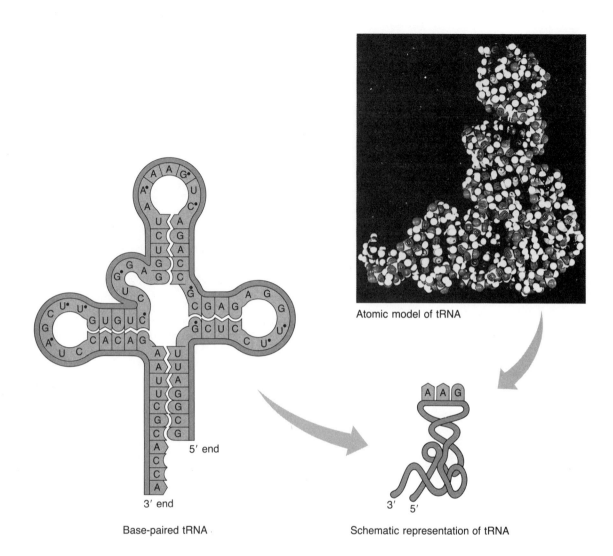

Atomic model of tRNA

Base-paired tRNA

Schematic representation of tRNA

Figure 2.13 Several representations of a transfer RNA chain folded into its characteristic shape. The bases with dots are commonly chemically modified in transfer RNA.

the extent of sequence relatedness between an isolated RNA and DNA can be estimated by measuring their ability to form a RNA–DNA double helix. Also, a DNA segment can be used as a means to detect the presence of a complementary RNA in a mixture of RNAs (and vice versa).

Figure 2.14 illustrates how the extent of complementarity between the nucleotide sequences in DNA and RNA chains can be estimated. First, the DNA double helix is unwound (denatured) by heating. RNA, which is here represented in color, is added to the solution of DNA and the temperature

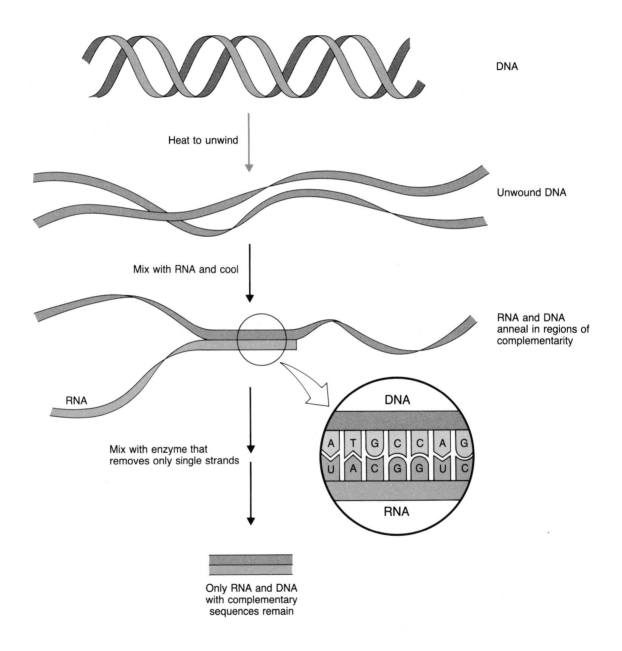

DNA

Heat to unwind

Unwound DNA

Mix with RNA and cool

RNA and DNA
anneal in regions of
complementarity

RNA

Mix with enzyme that
removes only single strands

DNA

A	T	G	C	C	A	G
U	A	C	G	G	U	C

RNA

Only RNA and DNA
with complementary
sequences remain

Figure 2.14   Measuring the extent of complementarity between DNA and RNA molecules.

is lowered. During annealing, base pairs form between complementary regions on the RNA and DNA. When unpaired, noncomplementary regions are removed with a special enzyme, the amount of remaining double helix indicates the extent of complementarity.

Proteins

PROTEINS are the principal determinants of an organism's characteristics. They constitute the enzymatic machinery for the synthesis of DNA and RNA, for the energetic and synthetic activities of all cells, for the regulatory elements that coordinate these activities, and even for their own synthesis. Proteins also contribute many of the structural elements that determine the shape and motility of cells and organisms. In short, organisms are what they are because of the array of proteins they manufacture.

At about the same time as the structure of the double helix was discovered, the final steps required to understand the basic structure of proteins were taken. Proteins consist of one or more chains that are, like nucleic acids, long polymers; these chains are called **polypeptides** (for a reason soon to be explained). Just as nucleic acids are polymers of nucleotides, polypeptides are linear polymers of **amino acids** (Figure 2.15). There are 20 different amino acids commonly found in the proteins of all living things. All these amino acids have in common one structural group of atoms (grey) and each amino acid has one unique chemical grouping called a side chain (here shown in color). A polypeptide chain is held together by strong chemical bonds between the carbon atom in the $C\!\!\diagdown\genfrac{}{}{0pt}{}{=O}{OH}$ (carboxyl) group on one amino acid and the nitrogen atom in the $-NH_2$ (amino) group on the neighboring amino acid. Such bonds are termed **peptide bonds**, thus the term "polypeptide" (Figure 2.16). Like nucleic acids, polypeptide chains have a direction. One end contains a free amino group, the amino terminus, and the other a free carboxyl group, the carboxyl terminus. The amino acid at the amino terminus is usually numbered as amino acid 1.

Polypeptides come in many sizes, generally between 50 and several thousand amino acids in length. But regardless of length, organism, or cell type, polypeptides—and thus all proteins—are constructed from the same repertoire of only 20 different amino acids. Each polypeptide possesses a unique sequence of amino acids along its length (Figure 2.17). This unique

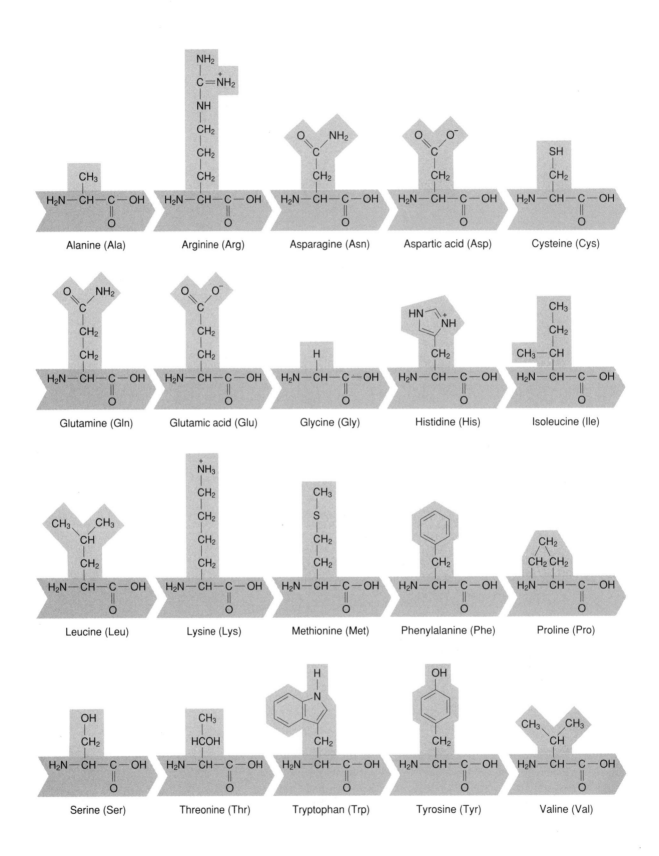

Figure 2.15 (*facing page*) Chemical structures of the naturally occurring amino acids. In many of the following figures, amino acids are represented by the three-letter designations contained within grey arrows.

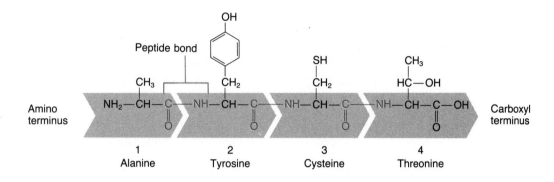

Figure 2.16 Peptide bonds between amino acids in a tetrapeptide (four amino acids).

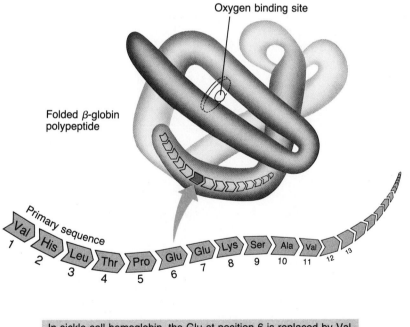

In sickle-cell hemoglobin, the Glu at position 6 is replaced by Val

Figure 2.17 A portion of the primary structure of the β-globin polypeptide and its location in the folded, complete polypeptide. Also shown is the amino acid that is altered in the β-globin polypeptide in sickle-cell disease.

amino acid sequence, called the **primary structure**, determines how the polypeptide folds up into a characteristic shape. The shape is determined by several factors, including the chemical properties of the side chains. Strictly speaking, the term protein refers to a **folded polypeptide.** Often, the proper three-dimensional form of a protein depends on the interaction of several polypeptide chains, which may be either identical or different in primary structure. Most important, it is this three-dimensional form that is responsible for the biological properties and functions of proteins and therefore their special role in cells.

Sickle-cell disease is one of many convincing examples of the dependence of normal protein function on the correct primary (and consequently the correct three-dimensional) structure. Hemoglobin is the protein in vertebrate red blood cells that carries oxygen from the lungs to all tissues. It contains four polypeptide chains, two each of two different polypeptide chains called by the Greek letters α ("alpha") and β ("beta"). Normal red blood cells appear round in a microscope; in people with sickle-cell disease, the same cells look like sickles (Plate II). Sickling results from a change in one amino acid among the 121 amino acids in the β chain (Figure 2.17). This simple change from one amino acid to another profoundly alters the normal interactions between hemoglobin molecules. When the red blood cells unload their oxygen in the tissues, the altered hemoglobin forms fibers, thereby distorting the red blood cell's shape. As a consequence, blood flow through capillaries and small veins is impeded or interrupted. The single amino acid replacement is caused by a change in a single DNA base pair in the gene for the β chain. In the most general sense, it is this relation between genes and protein structure that links an organism's properties to its genes.

Packaging DNA in Chromosomes

DNA probably never occurs free, in extended form, in living things. Rather, it interacts with many different proteins to form what is called chromatin. As a consequence of these diverse DNA–protein interactions, DNA is folded and condensed sometimes several thousandfold. Thus, as already mentioned, the 1.4-millimeter-long circular DNA of *E. coli* fits into a rodlike cell with a diameter of about 0.001 millimeter and a length of 0.002 millimeter, and about 2000 millimeters of DNA is confined to a nucleus about 0.005 millimeter in diameter in a nondividing human cell. The DNA in dividing cells is condensed even further, making it visible under a simple microscope as mitotic chromosomes.

Viruses

THIS chapter is mainly about the chemical structure of the genetic macromolecules. Because viruses are entities somewhere between complex aggregates of macromolecules and actual living organisms, this seems to be the most appropriate place to introduce them. More about the biology of viruses is found in Chapter 10.

Viruses are most commonly thought of as agents of human disease. Farmers and gardeners know that plants too are subject to viral infections. In fact, viruses infect most living species, including bacteria. In general, any particular virus can usually infect only one or a few related species. Bacterial viruses, called **bacteriophages** or **phages** for short (from the Greek *phagein*, "to eat"), do not infect plants or animals, and human viruses (such as those that cause measles, small pox, and AIDS) have no effect on bacteria. Typically, each species can be infected by a variety of distinctive viruses. The bacterium *E. coli*, for example, is a host for many different phages. In a way, viruses constitute a whole underworld of nature.

The genetic information in a virus, the virus's genome, may be either DNA or RNA. The genome is surrounded by a protein coat, which, like most biological entities, has a defined shape. *E. coli* phages T4 and lambda (both shown in Figure 2.18) each have a "head" and a "tail." The micrographs obscure the fact that the heads and tails are three-dimensional. In fact, the heads have 20 faces of nearly identical size. Viral genomes contain genes, and mutations in these genes produce alleles with varying consequences to the virus. The critical difference between a virus and a true living organism is a matter of autonomy. A virus may be viable on its own, but it cannot reproduce. To make more of its kind, a virus must enter a "host" cell, where it co-opts the cell's machinery for transcription, translation, and replication to express and reproduce its own genes. The infectious process, which usually leads to the release of new viruses that go on to infect new cells, may kill the host cell. Some viruses, however, multiply and are released without killing the host cell.

The genetic information of many viruses is contained in DNA, which is replicated, repaired, rearranged, and expressed in virtually the same way as a cell's DNA. The genomes of DNA viruses may be linear or circular double helices. Some of them are even linear or circular single strands of DNA. Other viruses have RNA genomes. Some RNA genomes replicate by copying the parental RNA genome (as a template) into new RNAs. The new RNA is then incorporated into progeny virus. In addition, there is a class of RNA viruses called **retroviruses**, whose RNA genomes begin their

a

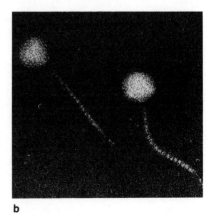

b

Figure 2.18 Two bacterial viruses: *a*, T4 phage, magnified 200,000 times; *b*, lambda phage, magnified 168,000 times.

reproductive cycles by a reverse transcription to form a complementary, single-stranded DNA "copy" of the RNA (Figure 2.1) and then a corresponding double-stranded DNA. The DNA is then inserted into the host cell's chromosome by recombination and becomes a permanent part of the host cell's genome; then it replicates in concert with the host's DNA. The genes in the viral DNA get transcribed and translated, producing progeny viral RNA genomes and virus-encoded proteins, and from these components new viruses are formed. They can exit the cell, enter a new cell, and reinitiate the virus's life cycle. One of the most infamous of such retroviruses is the human immunodeficiency virus that causes AIDS.

Viral genomes, like those of their host cells, may undergo recombination. Recombination may occur between two viral genomes within a cell and also between viral genomes and cellular DNA. These phenomena facilitated the mapping of bacterial and phage genes and led to an understanding of the overall organization of these genomes, which will be discussed in Chapter 4.

CHAPTER 3

Translating Genes into Traits

THE STORY of sickle-cell disease reveals the basic relation between gene structure and biological function. By the early 1940s, it was known that the disease is genetically determined and inherited according to the Mendelian rules for a single gene. Later, in the 1950s, it was learned that the mutation causing sickle-cell disease changed one amino acid in one of the polypeptide chains of hemoglobin. This finding supported the notion that genes specify the amino acid sequence of proteins, and it was consistent with the earlier conclusions relating a single gene to a single polypeptide. However, a more sophisticated insight into the relation between a gene's DNA sequence and the structure of the protein it encodes required a more amenable experimental organism than the human. For that purpose, researchers focused on the genes and proteins of prokaryotes, mainly bacteria of the species *E. coli*. The virtue of this system was the large number of mutants that were known to affect proteins needed for bacterial growth. Moreover, the analysis of the DNA, RNA, and protein structures affected by various mutations was considerably easier with bacteria than with more complex organisms.

Significant advances came from analyzing multiple mutations in an *E. coli* gene responsible for the formation of an enzyme required for the synthesis of the amino acid tryptophan. Genetic analysis was used to determine the relative linear order of the individual mutations in the *E. coli* gene, and chemical analysis of the altered enzymes' amino acid sequences was used to pinpoint the changes. In all cases, the linear order of the mutations along the gene were the same as the linear order of the amino acid changes in the protein. Because the DNA constituting the gene contains a specific linear sequence of nucleotides and the corresponding protein contains an equally specific linear sequence of amino acids, it was logical to conclude that the linear order of the four nucleotides in a stretch of DNA specifies the linear order of the 20 different amino acids along a polypeptide chain. The implied informational relation between nucleotide and amino acid sequences was called the **genetic code**. The challenge was to learn the code and how the message encoded in the gene's nucleotide sequence is translated into protein. Unexpectedly, the form and makeup of the code proved to be quite straightforward and elegant, and virtually the same in all forms of life. Moreover, the general rules for expressing genetic messages encoded in DNA are also universal. To understand the form of the code and how it was deciphered, we need first to consider the way genetic messages are translated into proteins.

DNA Begets RNA

THE EXPRESSION of all cellular genes begins with **transcription**, the process by which the nucleotide sequence in a gene's DNA is copied into an RNA chain (Figure 2.1). Transcription is carried out by **RNA polymerases**. These enzymes assemble RNA chains from individual nucleotides, the order of ribonucleotides being specified by base pairing to the sequence of deoxynucleotides in the DNA chain used as a template (Figure 3.1). Thus, a DNA strand serves as a template for RNA synthesis, in that an A, G, T, or C in the DNA template strand orders a U, C, A, or G, respectively, in the growing RNA chain. During transcription, an RNA polymerase molecule binds to discrete DNA sequences that define the beginning of each gene (**promoters**). At such promoter sequences, the enzyme separates the two strands of the DNA and selects one strand as the template for copying. Each nucleotide to be added to the RNA chain is determined by complementary base-pairing with the successive nucleotides in the selected DNA strand. As the RNA polymerase moves from the beginning to the end of the gene's coding sequence, each properly matched nucleotide is added to the growing end of the RNA chain. The newly made RNA strand has the same nucleotide sequence (except that U replaces T) and the same direction as one of the two DNA strands. Specific nucleotide sequences at the end of the gene signal the termination of transcription and trigger release of the completed RNA. Note that the DNA-directed assembly of RNA chains resembles DNA-directed DNA synthesis during DNA replication (Figure 2.10), the principal differences being that only one of the two DNA strands is copied into RNA, and that ribonucleotides are used to make RNA while deoxynucleotides are used to make DNA.

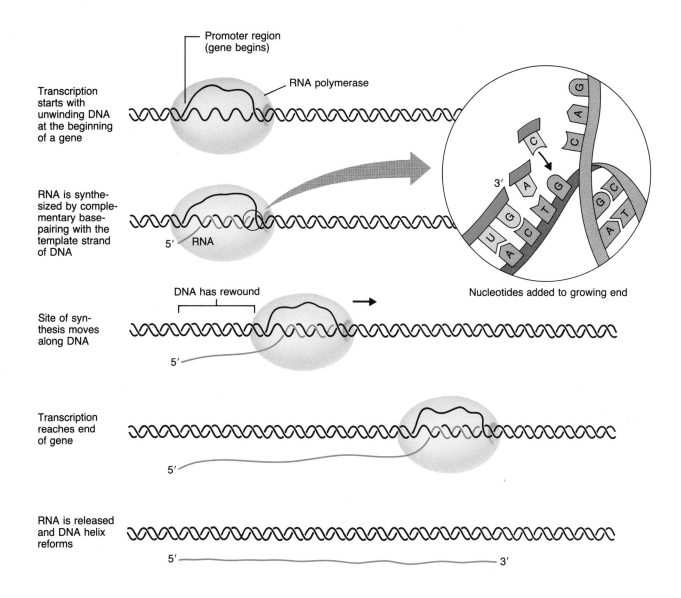

Figure 3.1 The basic features of transcription.

Transcription generates RNA molecules with different functions (Table 2.1, p. 50). Genes that specify the amino acid sequences of proteins are transcribed into **messenger RNAs (mRNAs)**. Some genes are transcribed into other kinds of RNAs, the most abundant being the **ribosomal RNAs (rRNAs)** and **transfer RNAs (tRNAs)**. The rRNAs and tRNAs do not encode proteins; rather, they are parts of the machinery that translates mRNAs into proteins. In prokaryotic organisms, a single RNA polymerase accounts for the production of all types of RNA. But eukaryotes use three distinctive kinds of RNA polymerase to make the three types of RNAs, providing these organisms with more refined mechanisms for regulating the production of the different RNAs.

The Genetic Code

BREAKING the genetic code was one of the monumental accomplishments in biology. Theoretical work and genetic experiments dominated the early attempts to learn the nature of the code. By the middle of the 1950s there were strong indications that there was a linear correspondence between the nucleotides in a gene's DNA and the amino acids in the protein it encodes (Figure 3.2). The mRNA transcribed from the gene's DNA was presumed to have the same linear correspondence. Genetic experiments had suggested that the coding unit for each amino acid most likely consisted of three consecutive nucleotides in RNA; such coding units were termed **codons**. Moreover, consecutive codons were presumed to code for adjacent amino acids in the proteins. But which of the 64 possible triplets (any of four possible nucleotides taken in groups of three produces 64 possible combinations) specified each of the 20 amino acids was a mystery with no apparent solution in sight. It was thus a surprise when a solution followed quickly from an important methodological breakthrough combined with a crucial but accidental discovery.

Biochemical studies during the late 1950s revealed that extracts of broken *E. coli* cells could synthesize polypeptides from individual amino acids supplied to the mixture. Then, in 1961, it was discovered that adding RNA molecules containing only one kind of base, uracil, to the *E. coli* extracts led to the formation of a polypeptide containing only one amino acid, phenylalanine. This suggested that UUU is the codon for phenylalanine. Indeed, other RNA molecules containing only one of the other three bases were quickly shown to code for polypeptides containing only one

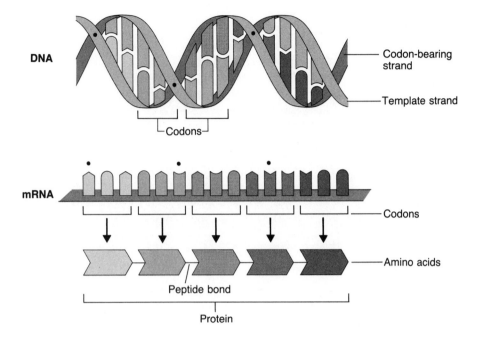

Figure 3.2 The linear correspondence between DNA base pairs, RNA nucleotides, and polypeptide amino acids.

particular amino acid. Thus, RNA with only adenines produced a polypeptide containing only the amino acid lysine (codon AAA). During the next few years, a variety of synthetic RNA molecules containing randomly arranged base sequences allowed deductions to be made about the composition of codons. This approach was improved when RNA molecules with defined nucleotide sequences were made and were shown to promote the formation of proteins with defined amino acid sequences. Subsequently, the 64 possible triplets were synthesized, and each was shown to specify one of the 20 amino acids or to provide punctuation signals in the coding sequence.

By 1964, the entire genetic code was known. Each codon comprises three adjacent nucleotides in a DNA chain or its mRNA copy (Figure 3.3). Sixty-one of the possible 64 triplets specify amino acids, and each encodes only one amino acid. One of these triplets, ATG in DNA or the equivalent AUG in RNA, has a dual function. It encodes the amino acid methionine, and it also marks the beginning of protein-coding stretches—the "**start**" **codon**. Three triplets—TAG (UAG), TAA (UAA), and TGA (UGA)—do not specify any amino acid; instead, each one of them signals the end of a protein-coding sequence—a "**stop**" **codon**.

64

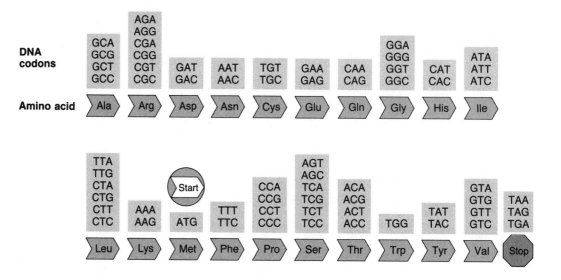

Figure 3.3 The genetic code. The codons shown for each amino acid are those for DNA. For RNA, the Ts are replaced by Us.

The code is said to be **degenerate** because more than one codon can specify the same amino acid; but, very importantly, the code is not ambiguous, because a particular codon never specifies more than one amino acid. With a knowledge of the genetic code, it is a straightforward exercise to translate on paper any DNA or RNA nucleotide sequence into its corresponding protein product. Usually, in any particular region of a long double helix, only one of the two DNA strands contains an informative array of codons that is translated into a protein. Generally, the sequence of the complementary strand is "nonsense," but occasionally that strand encodes part of another gene. Note, however, that it is the nonsense strand that is the template for mRNA synthesis. Synthesis produces the complement of the template—that is, the mRNA sequence.

The beginning of an mRNA chain corresponds to the position at which transcription of the DNA begins, and the end of the mRNA is where transcription ends. Because of the required direction of transcription, the mRNA is read in the 5' to 3' direction: 5' at the beginning and 3' at the end (Figure 3.4). The mRNA's nucleotide sequence is the same as that of the sense strand of the DNA. The transcription of a gene by RNA polymerase generally begins before the start of the protein-coding sequence and stops beyond the coding sequence's stop codon, so that mRNAs contain extra nucleotides at both ends.

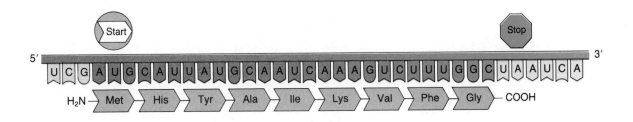

Figure 3.4   Decoding a messenger RNA sequence into a polypeptide.

The start of the protein-coding sequence is specified by an AUG codon (the start codon) located near the 5' end of the mRNA; thus, all proteins begin with methionine, the amino acid encoded by AUG. The end of the protein-coding sequence is signaled by a stop codon—UAG, UAA, or UGA—located close to the 3' end of the mRNA. The last amino acid of the polypeptide chain is therefore specified by the codon immediately preceding the stop codon.

How does each amino acid pair with its appropriate codons? There is no known chemical basis for a direct matchup. This is achieved, however, by an enzyme that attaches each amino acid to a special transfer RNA (tRNA). Each different tRNA has a triplet nucleotide sequence (an **anticodon**) that is complementary to the amino acid's coding triplet on mRNA. Base-pairing between the codon in the mRNA and the anticodon in the tRNA positions each amino acid at its proper codon and facilitates the joining of the amino acids to the growing end of the polypeptide chain. We shall see shortly how the tRNA bearing its amino acid is actually used during protein assembly.

RNA Begets Proteins

THE PROCESS by which the sequence of codons in an mRNA is translated into a polypeptide chain is complex, and it involves a very large number of repetitive steps. Physically, it is carried out by cytoplasmic structures that appear, under the electron microscope, as small particles (Figure 3.5). Called **ribosomes**, each of these particles itself consists of more than 50 different proteins and three or four different kinds of rRNA molecules (Table 2.1, p. 50). Ribosomes, together with tRNAs, constitute the machinery for converting nucleotide sequences in mRNAs into amino acid sequences in proteins. But, as we shall see, a variety of enzymes and other proteins are also needed.

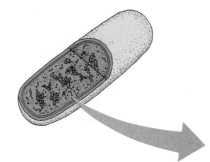

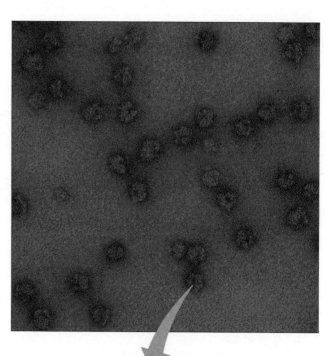

Figure 3.5   Ribosomes.

Note that there are three possible consecutive series of three nucleo-tides, or translation frames, in an mRNA, depending on which triplet is chosen as the first codon (Figure 3.6). In this example, as in most cases, two frames (labeled *B* and *C*) are interrupted by stop codons and cannot be translated. Only frame *A* is "open" throughout.

How is the correct translation or reading frame selected? This is ac-complished when a ribosome, together with a special tRNA carrying a methionine molecule and whose anticodon can base-pair with AUG, attaches to the mRNA at the start position's AUG. There are actually two tRNAs that can base-pair with AUG: One functions to start all protein chains with

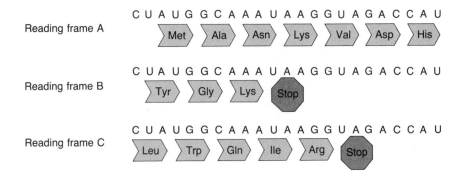

Reading frame A

C U A U G G C A A A U A A G G U A G A C C A U
Met Ala Asn Lys Val Asp His

Reading frame B

C U A U G G C A A A U A A G G U A G A C C A U
Tyr Gly Lys Stop

Reading frame C

C U A U G G C A A A U A A G G U A G A C C A U
Leu Trp Gln Ile Arg Stop

Figure 3.6   Alternate reading frames in a messenger RNA.

methionine, and the other serves to introduce methionine in response to AUG codons that appear in frame as the mRNA is being translated.

The special methionine-bearing initiator tRNA, along with the ribosome, binds to the start position's AUG codon (Figure 3.7). The next amino acid in the polypeptide chain is brought to the ribosome's translation site by a tRNA whose anticodon matches the second codon of the mRNA. Then the first peptide bond is made between the methionine and the second amino acid: The polypeptide chain has begun. As each codon is translated, one amino acid is added to the growing polypeptide chain. This process is repeated until all the codons in the coding sequence have been translated. The completed polypeptide chain is finally released when the translation apparatus reaches the translation termination signal, one of the three stop codons (UAA, UAG, or UGA).

An interesting and important feature of translation is that several ribosomes can translate an mRNA sequence simultaneously (Figure 3.8). Thus, after a ribosome has begun translating an mRNA, it moves away from the AUG that serves as the initiator codon, and a second ribosome can then initiate translation from the same start codon. After the second ribosome moves on, a third and fourth ribosome successively engage the mRNA chain at the same start position and initiate the assembly of additional polypeptide chains. Somewhat later, the first ribosome has completed the assembly of the polypeptide product and has released it. Concomitant with release of the completed polypeptide, the ribosome itself is released from the mRNA. Other ribosomes are nearing completion of the polypeptides they are making. In this way, many identical polypeptides can be read from a single mRNA strand in a very short time. The new polypeptides may

68

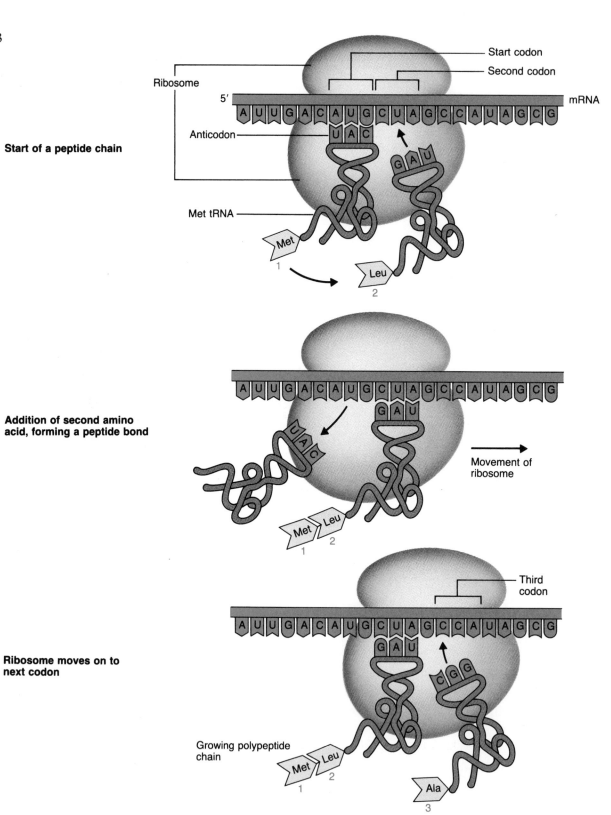

Figure 3.7 Initiating translation of a messenger RNA.

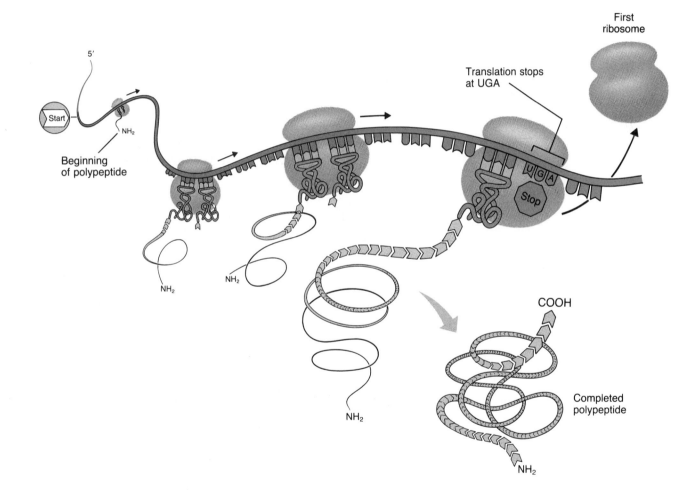

Figure 3.8 A messenger RNA is simultaneously translated by multiple ribosomes.

begin to fold into their active structures even before they are complete. Within a cell, ribosomes can be found all along mRNAs, each one's location revealing how much of the message it has already translated.

The two important directional rules of translation are as follows: Translation proceeds from 5′ to 3′ on the mRNA, and the protein grows from the amino end to the carboxyl end. It might help to recall that direction is also important in reading a sentence. Different languages print texts in different directions. Genetics has adopted conventions consistent with European languages. The beginning of the "sentence"—the 5′ end of mRNA and the amino end of the polypeptide—is at the left. The finish—the 3′ end of mRNA and the carboxyl end of the polypeptide—is at the right. By convention, therefore, the message is read from left to right.

Switching Genes On and Off

LEARNING how genes are transcribed and translated is only one aspect of the study of gene expression. Another aspect is the way in which the expression of a gene is regulated—that is, how the rate and amount of a gene's expression is determined under a wide variety of circumstances. Not surprisingly, progress in understanding the mechanics of transcription and translation illuminated the subject of gene regulation. Genetic studies with bacteria provided the tools and ideas for understanding how gene expression is regulated. Under certain conditions, many bacterial genes are not expressed at all, while the rates of expression of others may differ by several hundredfold. A change in conditions can trigger the activation of previously silent genes and the silencing of active genes. This capability provides bacterial cells with a broad-ranging flexibility to adjust to changing external environments—for example, to adapt to changing supplies of food and oxygen.

Gene expression is most frequently regulated at the level of transcription—that is, of mRNA production. Generally, initiation of transcription is the regulated event. Whether transcription of a particular gene is able to take place or not is determined by proteins that bind transiently to special DNA sequences immediately preceding the gene. Interaction of these regulatory proteins with the DNA can either block or allow transcription. Accordingly, the proteins are called **repressors** or **activators**, respectively. To appreciate the precise workings of such regulatory mechanisms and the way in which they permit an *E. coli* cell to respond to its environment, it is best to look at a particular example.

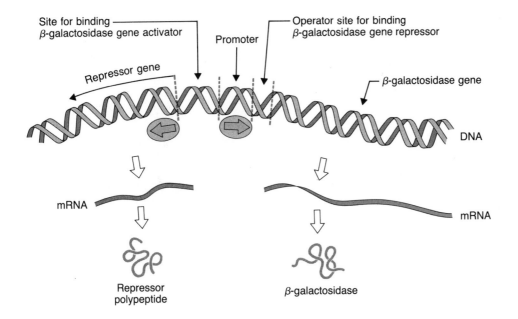

Figure 3.9 The DNA segments that regulate transcription of the β-galactosidase gene in *E. coli*.

The *E. coli* enzyme β-galactosidase (beta-galactosidase) breaks down lactose (milk sugar) into two simpler sugars, glucose and galactose. If *E. coli* is grown in the presence of glucose, its preferred nutrient, it does not synthesize β-galactosidase; but the enzyme is produced if lactose is the only sugar available. Several DNA sequences that precede the 5′ end of the β-galactosidase coding region on *E. coli* DNA serve to regulate transcription (Figure 3.9). The RNA polymerase enzyme that will transcribe the gene binds to one, the **promoter**. A second sequence, the **operator**, lies between the promoter and the start of the β-galactosidase coding sequence. The operator sequence interacts with another protein, the **repressor**. Binding of the repressor to the operator prevents RNA polymerase from initiating transcription.

If lactose is supplied to the *E. coli*, the sugar binds to the repressor protein, thereby altering the repressor's shape and preventing it from binding to the DNA (Figure 3.10). This loss of operator-binding activity now allows (1) the RNA polymerase to transcribe the gene into mRNA, (2) the β-galactosidase to be synthesized, and (3) lactose to be utilized as an energy source for growth. Note that, as in the case of the defective hemoglobin in sickle-cell disease, the particular structure of a protein is crucial for its

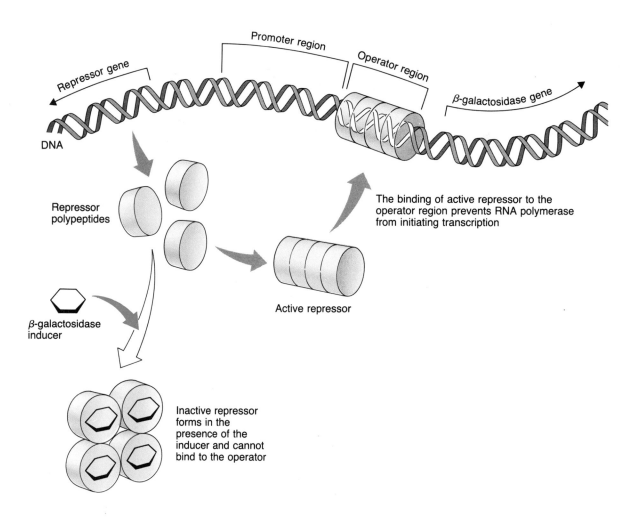

Figure 3.10 Alternate forms of the repressor, in the absence or presence of inducer, inhibit or permit, respectively, transcription of the β-galactosidase gene.

biological activity. It is important to realize that the β-galactosidase gene repressor specifically recognizes and binds to that gene's operator sequence. Because of the length and organization of the operator sequence, it is unlikely to occur anywhere else in the *E. coli* genome. Consequently, the repressor will not bind to other DNA sequences and, therefore, the repressor regulates only the β-galactosidase gene.

Besides the negative control provided by the repressor–operator interactions, β-galactosidase gene expression is also under positive control. Thus, transcription of the β-galactosidase gene can start only if a specific activator is present (Figure 3.11). The activator is also a protein that functions only when it is joined with a special small molecule. The special small molecule appears in the cell when there is no glucose available to the bacterium. In a sense, the small molecule is a starvation signal; under these circumstances, it forms a complex with the activator protein that binds to a short DNA segment near the promoter–operator region. When the complex is bound, it enhances the ability of RNA polymerase to transcribe the β-galactosidase gene. Thus, the expression of the β-galactosidase gene depends on two environmental conditions—the absence of glucose and the presence of lactose.

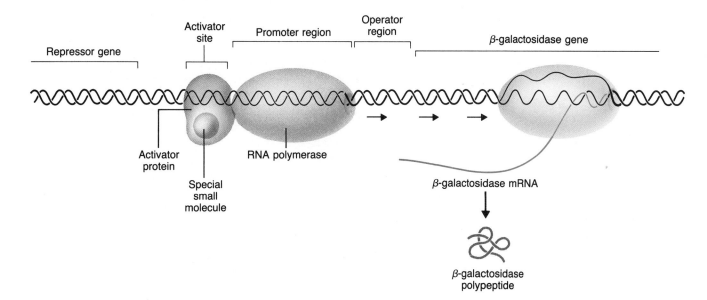

Figure 3.11   Transcription of the β-galactosidase gene requires an activator.

Transcription of most bacterial genes is subject to complex regulation of this sort, but repressor and activator proteins are not the sole means of regulating gene transcription. In some instances, the protein product of a gene's expression is itself the regulator of that gene's transcription, providing a feedback regulatory system. The production of mRNA can also be regulated, either by controlling the rate at which RNA chains grow rather than the rate of their initiation, or by altering the likelihood that transcription will continue through the entire gene or be terminated at control points within the gene.

Gene expression can also be regulated during the translation of mRNA into polypeptide. Here too, control is most often exercised at the initation step, the reading of the first codon, although later steps in polypeptide chain synthesis can also be regulated. Additional regulatory events can occur during conversion of a completed polypeptide into a functional protein, because many polypeptides are not functional unless modified by the addition of special chemical groups. For example, the iron-containing structure in hemoglobin that is responsible for transporting oxygen is added to the protein only after the two different polypeptide chains are made and assembled. Each such modification is carried out by one or more enzymes whose abundance or level of activity can itself be modulated. Moreover, many proteins must be transported to a particular cellular location in order to function properly. For example, in eukaryotic cells, the proteins required to form a ribosome must be transported from the cytoplasm where they are made, to the nucleoli, where ribosomes are assembled. Other proteins must be delivered to the cell membrane or, like hormones, to the outside of the cell. Consequently, controlling the transport of proteins to their proper sites also regulates the levels of active proteins. Thus, numerous diverse and complex mechanisms cooperate to regulate the amount of a gene's functional product. Often, the processes are specialized for a particular gene, for the cell's physiological condition, or for the external environment.

Confronting Complex Genomes

A REMARKABLE interplay between genetics and biochemistry helped unravel the complexities of the genetic mechanisms of bacteria. As described in the next chapter, the combination of genetic and biochemical techniques also led to the isolation of a few bacterial genes. This paved the way for even more detailed analysis of gene transcription and translation. The logic governing the regulation of gene expression through molecular interactions of specific proteins (such as activators and repressors) with corresponding regulatory nucleotide sequences on DNA or mRNA is now known to be widespread.

By comparison with the energetic and fruitful analysis of bacterial (prokaryotic) genomes during the 1950s and 1960s, progress toward understanding eukaryotic genomes was frustratingly slow. The enormous size of most eukaryotic genomes was itself a formidable impediment to unraveling their structure, organization, and function (Table 3.1). Compounding the problem was the difficulty of genetic analysis. Although sophisticated genetic maps of mutations existed for those few eukaryotes whose genetic systems were experimentally manipulable, such as brewer's yeast, pink bread mold, and the fruit fly, genetic maps for higher organisms, such as mice and humans, were primitive by comparison. Ignorance of the molecular features of the genetic systems of higher organisms was even more profound. The structure of eukaryotic genes and their organization within chromosomal DNA were virtually complete mysteries, and the abundance of highly repeated DNA sequences in most eukaryotes posed an additional enigma. Without a more thorough analysis of the molecular anatomy of eukaryotic genomes, it was impossible to proceed.

Biochemical experiments on eukaryotic gene expression and regulation were also stymied by the inaccessibility of structural information about cellular genes. It was clear that the nuclear DNA of eukaryotes is transcribed into RNA and that mRNAs are translated into proteins by a process mediated by ribosomes and tRNAs, much like the process in prokaryotes and using the same genetic code. It was also known that, whereas transcription occurs in the nucleus, the resulting mRNAs needed to exit the nucleus in order to be translated: Eukaryotic ribosomes, and thus protein synthesis, occur only in the cytoplasm.

The nature of the eukaryotic transcription process itself and the subsequent fate of the RNA products (transcripts) were particularly puzzling. In many eukaryotes, less than 10 percent of the RNA made in the nucleus ends up in the cytoplasm as mRNA, rRNA, and tRNA. Some short RNA chains are sequestered in stable form in the nucleus, within particulate structures that also contain proteins, but most RNA in the nucleus is rapidly destroyed there. The origin and function of the rapidly synthesized and promptly degraded RNA were vexing issues. The question of mRNA formation was also confounded by the fact that the structure of eukaryotic mRNAs differ from their prokaryotic counterparts. Eukaryotic mRNAs have modifications at both ends of the chain—a so-called "cap" at the beginning

Table 3.1 *Examples of Genome Sizes*

Species	Approximate size of haploid genome (millions of base pairs)	Haploid number of chromosomes
VIRUS		
Phage lambda	0.05	1
PROKARYOTE		
Escherichia coli	4	1
EUKARYOTIC CELLS		
Yeast (*Saccharomyces cerevisiae*)	14	16
Trypanosome (*Trypanosoma brucei*)	80	unknown
Nematode (*Caenorhabditis elegans*)	80	11 or 12
Fruit fly (*Drosophila melanogaster*)	170	4
Toad (*Xenopus laevis*)	3000	18
Chicken (*Gallus domesticus*)	1200	39
Mouse (*Mus musculus*)	3000	20
Human (*Homo sapiens*)	3000	23
Corn (*Zea mays*)	5000	10
Mouse-ear cress (*Arabidopsis thaliana*)	70	5

Recall that the eukaryotic chromosomes come in pairs of almost identical homologous chromosomes. The number given here is for the number of pairs. The actual number of chromosomes is two times the number given here. Similarly, the haploid genome size is given. The actual number of base pairs of DNA in each eukaryotic cell is twice what is shown.

(5' end) and a "poly-A tail" at the 3' end. This implied that eukaryotic mRNAs underwent modifications after they were transcribed. Why, where, and how do the modifications occur? What is their significance? What is the pathway for converting raw RNA transcripts into such "mature" mRNAs? How does mRNA travel from the nucleus to the cytoplasm? Further, why is the rate of transcription of the same gene different in different cell types of the same organism? In what ways do the regulatory mechanisms of gene expression differ between eukaryotes and prokaryotes? In the absence of hard-core molecular genetic information and the methodology to obtain it, these vital questions about eukaryotes were unanswerable.

Therefore, it was not surprising that, as more and more was learned about how genetic information is organized and expressed in prokaryotes, the ignorance about eukaryotes became increasingly frustrating. Imagine trying to understand, for example, how culture influences political systems, but being confined in one's investigations to a single rather small country, with no books, periodicals, or information of any kind about the rest of the world. Knowing that there must be a wealth of different kinds of cultures and political systems out there but not knowing how different they are, one's own thoughts and ideas could hardly be satisfying. Some scientists tried to resolve the problem by assuming that eukaryotic cells are fundamentally the same as prokaryotic cells, only more complex. This view was bolstered by the discovery of the universality of many biological molecules and mechanisms—DNA, the genetic code, transcription, translation, and the role of complementary base-pairing. But it ignored the hints summarized in the preceding paragraph, hints that foretold profound differences between the genetic chemistry of prokaryotes and that of eukaryotes.

What was needed was a general methodology that would facilitate the molecular analysis of eukaryotic genomes. Ideally, such a breakthrough would circumvent the difficulties, the expense, and the time required to carry out experimental breeding with most eukaryotes. It would permit the isolation of discrete genes and the determination of their molecular structures and their genomic organization. Such isolated genetic elements could then be used in biochemical experiments for the purpose of characterizing the transcriptional and translational mechanisms that govern their expression. That objective finally became a reality during the first half of the 1970s. Before considering these developments in later chapters, we will, in Chapter 4, examine the origins of the concepts and methodologies that paved the way for the new technology.

CHAPTER 4

Learning to Deal with Genes

A
S IS OFTEN the case in scientific advances, many seemingly un-
connected discoveries paved the way for the new genetic techniques
that are referred to collectively as **recombinant DNA** methodology. Among the discoveries that were essential to the development of
the new methods was that of bacterial **plasmids**, small DNA molecules
that are not part of the cell's chromosome. Another was the discovery of
restriction nucleases, enzymes that cut DNA molecules at specific nucleotide sequences and thereby permit the precise manipulation of DNA
molecules. Especially critical were experiments in bacterial genetics and,
most especially, the discovery that DNA molecules can be introduced into
bacterial cells, which can then express the newly acquired genes. The inserted
DNA may be from a bacterial virus (a **bacteriophage**, or **phage** for short),
from other bacterial cells, or from a plasmid.

The theoretical and experimental background for the recombinant DNA
methods are described in this chapter. Both classical genetics and biochemistry figure importantly and many new ideas are presented. Chapter 5
reviews some of this material from a different perspective and provides
further clarification.

Moving Genes Between Cells

S
EXUAL reproduction provides a way for the genetic information from
two different individuals to become mixed. In animals, this occurs following fertilization of an egg by a sperm; in plants, this is achieved by
pollination. Bacteria can acquire new genetic information in several ways.
These include processes by which cells acquire DNA molecules from the
surrounding medium, or by direct transfer from another cell, or after infection
by phage particles. Whatever the mode of entry into the recipient cell, the

acquired DNA may recombine either with homologous regions or with special sites in the recipient organism's DNA, thus becoming part of its genome. Alternatively, it may be maintained as a minichromosome, replicating autonomously in the cell and being passed on to daughter cells upon division.

Bacterial transformation refers to the alteration of a cell's genome following the uptake of DNA molecules from the culture medium (Figure 4.1). The first experimental introduction of new genes into bacteria was by this method. In fact, this discovery provided the proof that DNA is the carrier of genetic information. In practice, when a population of cells with a particular genetic disability (for example, an inability to synthesize the amino acid tryptophan, or to metabolize the sugar lactose, or to produce a particular cell-surface molecule) is exposed to DNA purified from cells that can perform that function, a few of the cells acquire the capability characteristic of the donor DNA. Such transformations are stable and heritable by the cells' offspring because the newly acquired functional gene becomes part of the genomic DNA of the recipient cells. Bacterial transformation by DNA proved to be only marginally useful for the early study of prokaryotic molecular genetics for several reasons that are not important here. However, the principle of transformation is now of major significance because it is an essential step in many recombinant DNA experiments; and the term "transformation" has been adopted generally, with prokaryotes and eukaryotes, to refer to the permanent alteration of a cell's genetic information and its effect on the cell's characteristics.

Conjugation refers to the direct transfer of DNA from one bacterial cell to another upon contact, as seen in Figure 4.2. Two *E. coli* cells, one capable of donating its chromosome to the other, make contact through a conjugation bridge composed of protein. DNA replication starts at a special position in the donor cell's chromosome, shown in the figure by an arrowhead. One molecule of newly replicated DNA is transferred to the recipient cell. The transfer stops when the bridge is disrupted—for example, by random motion of the cells, or by shaking the flask containing the cells. The longer the two cells are in contact before the bridge breaks, the greater the number of genes that are transferred. The transferred DNA can then replace the corresponding region in the recipient cell's genome by recombination. This process can be detected when, for example, the donor DNA provides a normal gene to replace a nonfunctional one in the recipient.

Bacterial geneticists were able to learn the order in which different genes were transferred relative to the duration of conjugation (Figure 4.3).

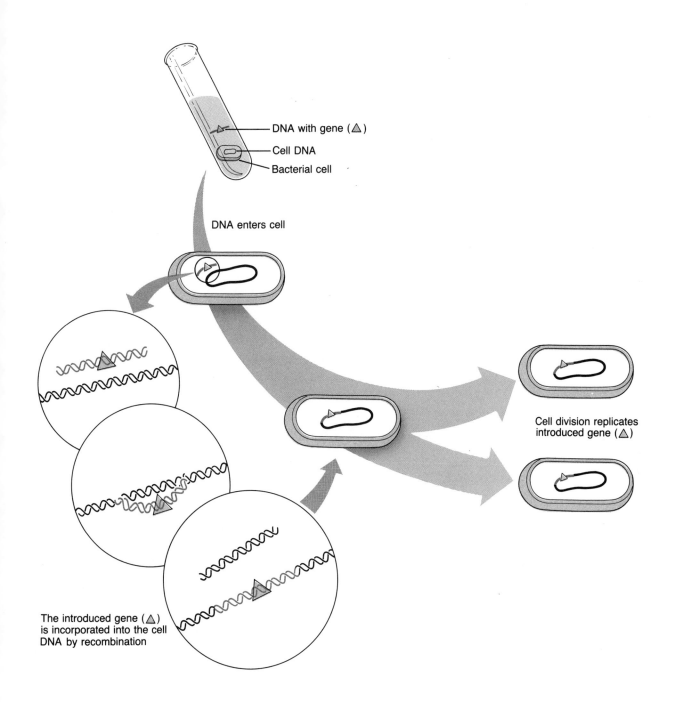

DNA with gene (△)

Cell DNA

Bacterial cell

DNA enters cell

Cell division replicates
introduced gene (△)

The introduced gene (△)
is incorporated into the cell
DNA by recombination

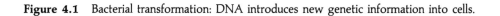

Figure 4.1 Bacterial transformation: DNA introduces new genetic information into cells.

82

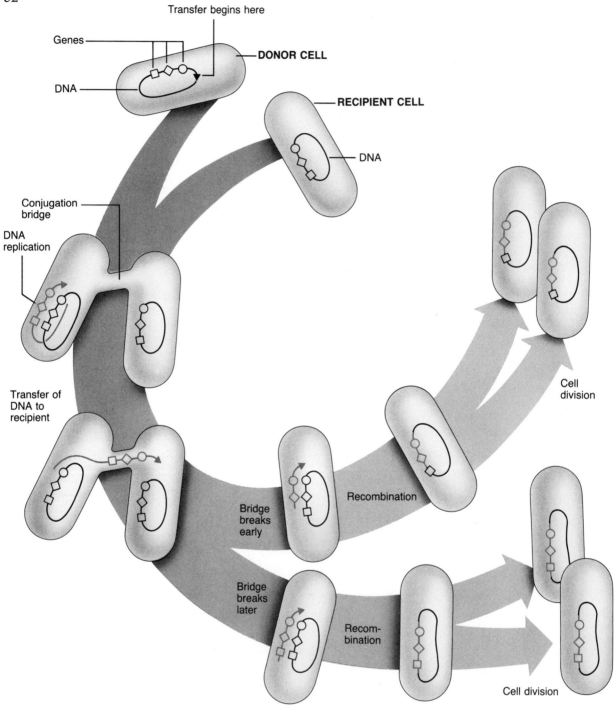

Figure 4.2 Bacterial conjugation: transfer of genetic information between bacterial cells.

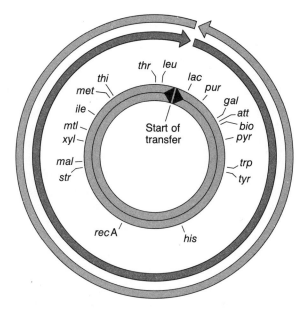

Figure 4.3 Circular genetic map of *E. coli* showing the locations of several representative genes. The arrows indicate the directions of gene transfer from two different donors during conjugation.

They discovered that the order followed different patterns, depending on which strain of *E. coli* was used as the donor. In one pattern (see inner circular arrow on Figure 4.3), for example, the order was, from early to late transfer,

leu, met, xyl, his, lac

(in which each three-letter abbreviation designates a different gene). In another pattern (see outside circular arrow on Figure 4.3), the order from early to late transfer was the opposite,

lac, his, xyl, met, leu.

This is what might be expected if the genes are lined up on a chromosome of linear double-helical DNA, and if transfer can begin at either end. However, conjugative transfer showed another peculiarity. Depending on the strain of the donor *E. coli*, transfer begins at different positions on the chromosome. Thus, among the strains that transfer in the first order shown above, some give the series *leu, met, xyl, his, lac*, while others transfer *met* first, followed by *xyl, his, lac*, and, much later, *leu*. The explanation of these results lies in the fact that the *E. coli* chromosome is a circular DNA double helix. It has no ends, and transfer can start from many positions on the circle and can proceed in either direction. In fact, it was these genetic experiments that first indicated that all the *E. coli* genes are on a single, circular DNA molecule. The order of genes on the *E. coli* chromosome, the **genetic map**, was constructed by comparing the length of time required for different genes to be transferred upon conjugation (Figure 4.3).

Infection of bacterial cells by phages begins when a phage attaches to a cell and injects its DNA into the cytoplasm (Figure 4.4). Enzymes encoded by both phage and cell genes work together to make many new copies of the phage DNA and proteins that are assembled into new phage particles.

In addition to their own DNA, phages can acquire cell DNA from the bacteria they infect and transfer that newly acquired DNA into other bacteria in the course of a subsequent infection. This phenomenon is called **transduction**. Two types of transduction have been characterized. In one type, the transducing phage acquires the foreign DNA during the course of an infection when the bacterial DNA is degraded into fragments about the size of the phage genome (about 50,000 base pairs) and the fragments are accidently packaged into the phage structure (Figure 4.4). Such transducing phages introduce the newly acquired donor-cell DNA into a recipient cell during subsequent infections. The introduced donor-cell DNA fragment replaces the corresponding segment of DNA in the DNA of the recipient cell and becomes a permanent part of that cell's genome. Each transducing phage generally contains only a single random fragment of the original donor chromosome. Moreover, there is a nearly equal probability that the fragment is derived from any particular portion of that genome. However, because the transduced DNA segments are quite large (50,000 base pairs is more than 1 percent of the 4 million base pairs on the *E. coli* chromosome), and any particular bacterial gene is rarely more than 1000 base pairs long, the recipient cell generally acquires a group of genes in a single transduction event. Thus, genes that are close to each other on the donor chromosome are frequently transduced together on a single fragment. Genes that are far apart are transduced independently. Measurements of the frequency with which genes are transduced together provides an estimate of the relative distance between genes that are relatively close together to begin with.

The second type of transduction is an attribute of phages whose normal infectious cycle can be, and frequently is, interrupted by the integration of the viral genome into a special location in the infected cell's chromosome (Figure 4.5). This is a special kind of recombination. Both phage and cell DNAs are cleaved at their special locations and then joined together. The

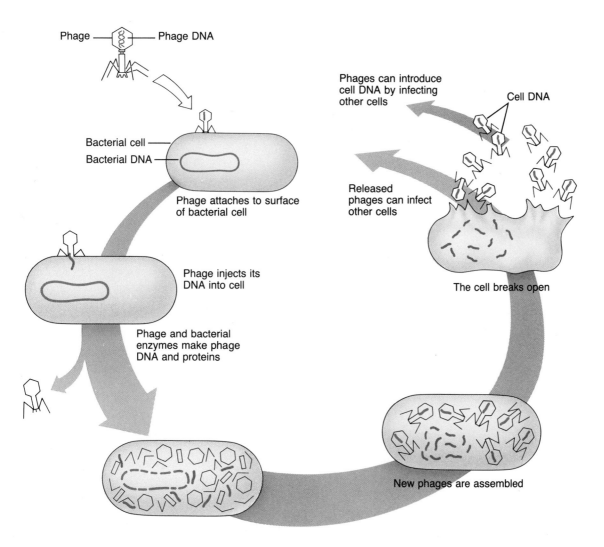

Figure 4.4 Events that occur when a phage infects a bacterial cell.

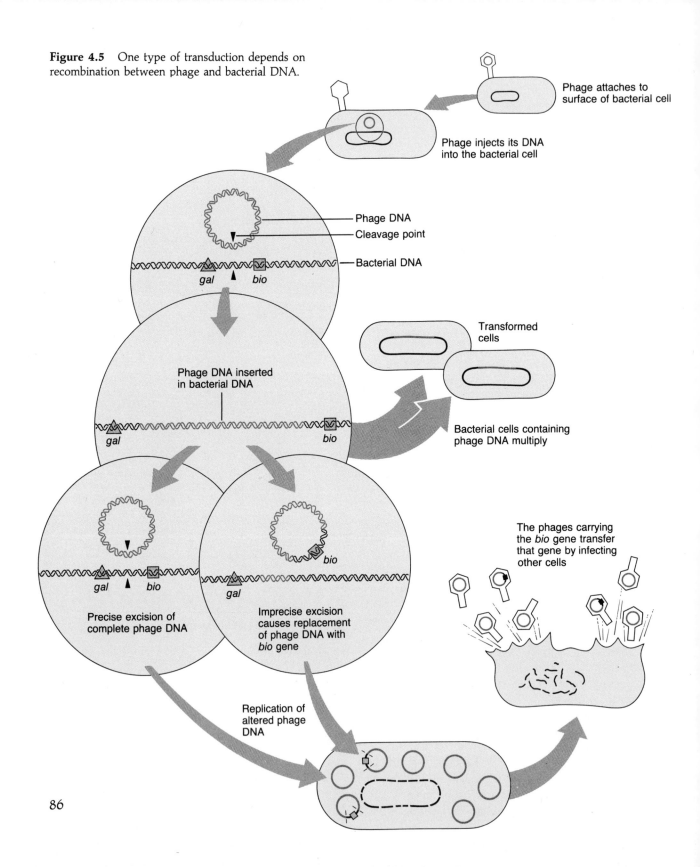

Figure 4.5 One type of transduction depends on recombination between phage and bacterial DNA.

Phage attaches to surface of bacterial cell

Phage injects its DNA into the bacterial cell

Phage DNA
Cleavage point
Bacterial DNA

gal   *bio*

Transformed cells

Phage DNA inserted in bacterial DNA

gal   *bio*

Bacterial cells containing phage DNA multiply

The phages carrying the *bio* gene transfer that gene by infecting other cells

gal   *bio*

gal   *bio*

Precise excision of complete phage DNA

Imprecise excision causes replacement of phage DNA with *bio* gene

Replication of altered phage DNA

breaks occur across both strands in each DNA, and the broken ends are then joined together by the action of the enzyme DNA ligase. The result is the insertion of the phage DNA into the bacterial chromosome. Bacteria containing such integrated phage DNA pass the phage genomes along to their daughter cells as a part of their own chromosomes. The cell and viral genomes replicate as one and maintain a mutually compatible existence. Integration of the phage genome into cellular DNA is a nonlethal alternative to the usual outcome of such infections—namely, the death of the infected cells and the production of new infectious virus.

Under certain environmental conditions, an integrated phage genome excises itself from the cell's chromosome. As a consequence, the phage DNA replicates, produces new infectious virus, and kills the cell. Usually, excision of the viral DNA is precise, and the resulting phage genome is exactly like that of the original virus (Figure 4.5, bottom left). Occasionally, however, excision of the phage DNA is imprecise, and cellular genes that are adjacent to the integrated phage genome are incorporated into the progeny phage genomes in place of some phage genes. Two such adjacent genes—*bio* and *gal*—are indicated in Figure 4.5, and, in this example, the *bio* gene is included in the imprecisely excised DNA. During the next cycle of infection, the cellular gene (in this case, *bio*) is introduced along with phage genes into the newly infected cells. Following insertion of the phage DNA into the genome of the recipient cell, the cell acquires, in addition to phage genes, genetic information that originated from the phage's previous host. Thus, the phage serves as a carrier, or **vector**, for transducing genes from one cell to another; only cellular genes that are close to the integration site of the phage DNA are transducible by this mode.

Because different phages integrate into distinct sites on a bacterial chromosome, they can transduce different chromosomal genes. Certain genetic tricks enable phages to acquire even those genes that are far away from their integration sites. Transducing phages can acquire mutant genes as easily as normal genes if the mutants are present in the original donor. Because transducing phages can be readily identified, grown, and purified, it is possible to obtain substantial amounts of their DNA. Therefore, normal or mutant forms of *E. coli* genes are recoverable in enriched form.

The enrichment of a bacterial gene that is afforded by its incorporation into the DNA of a transducing phage is dramatic. Consider, for example, the *E. coli* gene, *lacZ*, that encodes β-galactosidase. The β-galactosidase protein is composed of four identical polypeptide chains, each containing 1173 amino acids. The DNA sequence encoding the polypeptide is thus 3519 base pairs long (3 times 1173). Whereas *lacZ* represents about one-thousandth of the *E. coli* genome (3519 out of 4,000,000 base pairs), it is

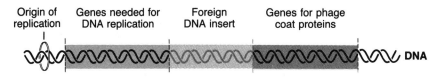

Figure 4.6   Essential features of a transducing phage.

about one-fifteenth of the genome of a transducing virus (3519 out of 50,000 base pairs). For this reason, it is much easier (about 100 times easier) to isolate the *lacZ* gene from the DNA of the transducing virus than from *E. coli* DNA. This simplification in the isolation of the *lacZ* gene aided in the identification of the DNA segments that regulate that gene's expression and the analysis of its nucleotide sequence. In a similar way, transducing phages that integrate at other locations in the bacterial genome were used for the isolation of other genes. The analysis of their nucleotide sequences identified their coding sequences and the regulatory signals that govern their expression. Moreover, because it was possible to transduce mutant as well as normal bacterial genes, the effect of various kinds of mutational alterations could be tested for their effects on that gene's expression and regulation.

Consider the special attributes of a phage genome that enable it to function as a viable transducing phage (Figure 4.6). First, because the genome must replicate following infection, the phage DNA must retain a site at which replication starts—an **origin of replication**—and the genes necessary to carry out DNA replication. Second, it must be able to acquire a segment of nonviral DNA—that is, the segment being transduced. The nonphage DNA may be inserted at any position in the phage genome, so long as it does not impair the capacity of the phage DNA to be replicated in the infected host cell or diminish the likelihood that it will be packaged into phage particles. Having become an integral part of the phage DNA, the transduced DNA segment is replicated along with the phage genome. Third, genes that encode the coat proteins of the phage must be present. Fourth, because the transducing phages obtained after an infection are usually mixed with phages containing unaltered phage DNA (Figure 4.5), there must be a way to separate the normal phages from those that carry the newly acquired DNA, and to identify the phages that have the particular gene of interest. Such separation is generally achieved by **cloning**, as described in the following paragraphs.

Cloning

IN ORDER to understand how the concepts of transduction were adapted for the recombinant DNA methods, we need to examine the significance of cloning. The word "**clone**" has entered the vernacular; it can be found in newspaper articles, in novels and in poems, and it is heard on radio and television. Often it denotes a single, perfect copy of something—a person, an animal, an idea—but, this is not the way in which biologists generally use the word.

A clone is a population of genetically identical organisms, cells, viruses, or DNA molecules. Such a population of individuals is derived from the reproduction of a single cell, virus, or DNA molecule. All the members of a clone—whether viruses, or cells, or molecules—are virtually identical to the virus, or cell, or molecule that initiated the production of the clone. (We say *virtually* because some new mutations are bound to occur during the generations required to produce the clone.) All the members of a clone are also identical to each other (with the same proviso). With viruses and cells, this means that their genomes have the same nucleotide sequence. Thus, cloning viruses and cells is a way to clone DNA molecules.

Cloning of viruses is initiated by infecting a single cell with a single virus particle. The virus multiplies and then the progeny infect new cells. In practice, many cycles of infection can be carried out without mixing the virus progeny with other types of virus. Such viral clones are seen as isolated clear areas ("plaques") on a layer ("lawn") of uninfected bacterial cells growing in a Petri dish (Figure 4.7). The clear areas form because the infected cells, surrounding the first infected cell, are all dead.

Cloning of cells is achieved if the cells in a mixture are permitted to multiply in isolation from one another (Figure 4.8). Clones of bacterial or mammalian cells are readily produced when cells are spread sparsely on a suitable growth medium in a dish. When they grow and multiply, each of them forms an isolated multicell colony. As a practical matter, cell colonies are grown either on the surface of, or within, or under a suitable semisolid material containing the nutrients. Usually the material is gelatinous agar, which is derived from seaweed.

Cloning is a means for obtaining a preparation containing a single kind of DNA molecule, be it from a virus or a cell, because each individual virus or cell in a clone has the same DNA. In principle, a single cell contains only a single genome, and, therefore, millions of cells provide millions of identical copies of the DNA genome for chemical manipulation. In the next section, we discuss how the principle of cloning is used to obtain pure preparations of experimentally restructured (recombinant) DNA molecules.

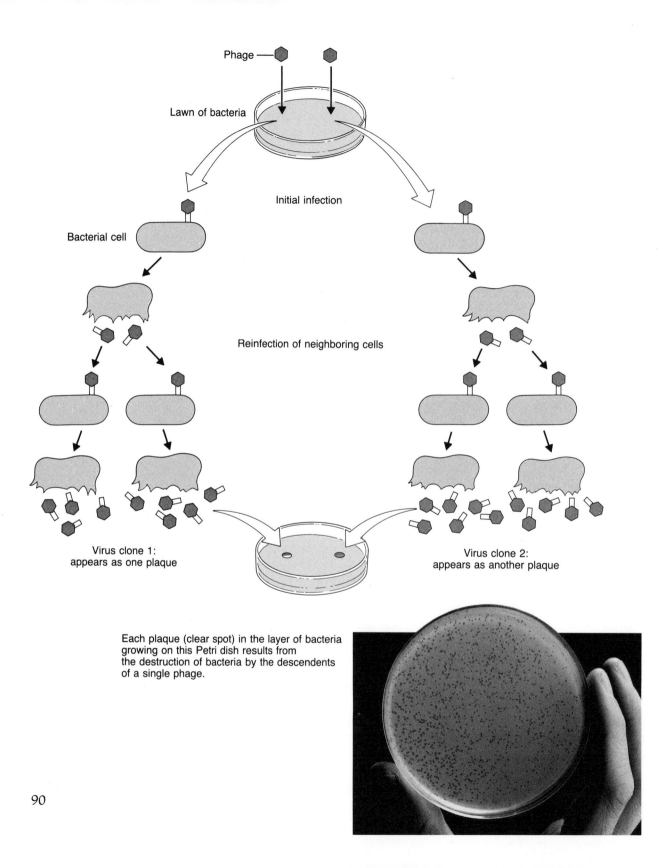

Phage

Lawn of bacteria

Initial infection

Bacterial cell

Reinfection of neighboring cells

Virus clone 1:
appears as one plaque

Virus clone 2:
appears as another plaque

Each plaque (clear spot) in the layer of bacteria growing on this Petri dish results from the destruction of bacteria by the descendents of a single phage.

Figure 4.7 (*facing page*) A clone of virus particles is a population of viruses derived from the multiplication of a single particle. The figure uses a bacterial virus — phage — as an example, but the principles are the same for plant and animal viruses.

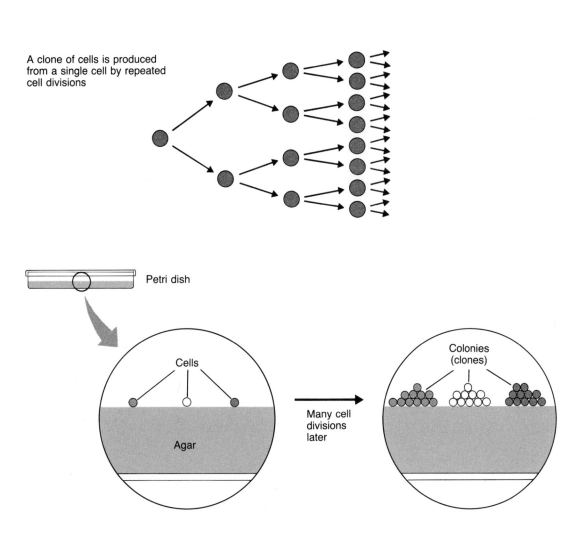

A clone of cells is produced from a single cell by repeated cell divisions

Petri dish

Cells

Agar

Many cell divisions later

Colonies (clones)

Figure 4.8 A clone of cells is a population of cells derived from a single cell by multiple cell divisions. The top half of the diagram illustrates how sequential cell divisions result in the formation of a clone. The bottom half shows how clones appear on a Petri dish.

Recombinant DNA

RECOMBINANT DNA methodology depends upon the same basic principle that underlies transduction. Foreign DNA segments can be carried into cells once they have been inserted into DNA molecules capable of entering and replicating in those cells. The DNA molecules that serve as vectors can be viral genomes or plasmids. Foreign DNA segments inserted into vector DNAs will hereafter be referred to as **DNA inserts**.

Scientists no longer need to rely on the relatively rare cellular processes that produce transducing phage genomes. They now produce them easily and at will in laboratory test tubes by joining DNA inserts to phage DNAs through the action of the enzyme DNA ligase (Figure 4.9). (DNA ligase was mentioned earlier for its ability to join cellular DNA together during replication.) These artificially produced recombinant DNA molecules are then introduced into cells in which they can be amplified through DNA replication.

The enormous potential of this methodology stems not simply from the construction, replication, and isolation of recombinant DNAs but also from our ability to clone individual molecules of recombinant DNA. Consider, for example, the consequences of joining a mixture of different DNA molecules to a population of identical DNA molecules from a phage vector (Figure 4.10). The mixture of different DNA molecules might be fragments

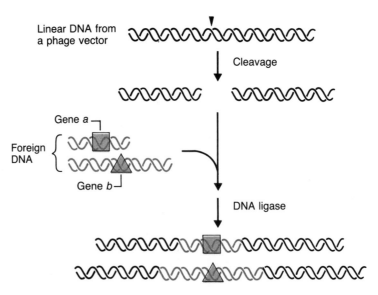

Figure 4.9 Joining foreign DNA fragments into vector DNA.

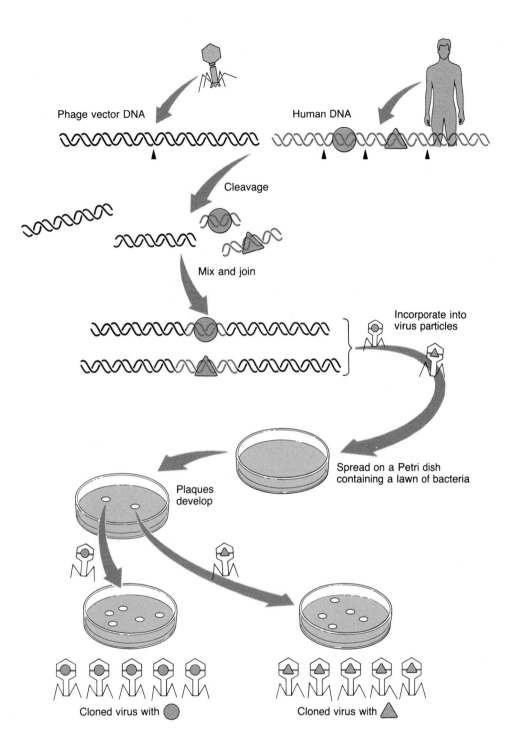

Figure 4.10 Molecular cloning of a mixture of foreign DNA inserts in a viral vector.

representative of the entire genome of some organism; it could, for instance, have been obtained from your own white blood cells. The diploid human genome has 6 billion base pairs (see Table 3.1). If the DNA were broken up into fragments, each 20,000 base pairs long, there would be, on the average, about 300,000 different fragments. Each fragment comes from a different part of the genome, has a different sequence of nucleotides, and contains different genes. The mixture of fragments is added to a solution containing millions of the identical vector DNAs. Only one human DNA fragment inserts into any one vector DNA. The solution now contains about 300,000 recombinant DNAs; each has a different fragment of human DNA inserted into an identical vector DNA. The solution also contains a lot of extra vector DNA molecules that did not receive any insert at all. All the DNA molecules in the solution are then packaged into phage particles and cloned as individual virus plaques. Each plaque contains either a unique recombinant DNA molecule (composed of the vector DNA and a single fragment from the original genome) or an unaltered vector DNA. Overall, the recombinant DNA technique makes it possible to recover substantial amounts of single DNA fragments from extremely complex mixtures of fragments produced from any organism's genome.

We pointed out earlier (p. 87) that the transfer of an *E. coli* gene from the bacterial genome to the genome of a transducing phage achieved approximately a 100-fold enrichment of that gene. This is a small advantage compared to what is achieved by molecular cloning of DNA fragments from complex organisms. For example, a mammalian gene containing 5000 base pairs is somewhat less than one-millionth of the mammalian genome (5000 base pairs out of 6,000,000,000 base pairs), but it is about one-tenth of the genome of a recombinant phage (50,000 base pairs). Thus, molecular cloning provides a way to divide even the largest and most complex genomes into discrete segments containing one or a few genes and to recover these in relatively pure form. This quite straightforward application of the principles and methods that helped scientists to analyze the molecular genetics of prokaryotes removed the barrier that had prevented similar analysis of eukaryotic genomes.

Novel Vectors

ONE OF the striking consequences of the use of antibiotics for the treatment of infectious disease was the emergence of strains of pathogenic (disease-causing) bacteria that resist antibiotics. The serious medical consequences of this problem stimulated research seeking to explain the phenomenon. Soon it became apparent that drug resistance is a relatively stable genetic trait. It can be transmitted from antibiotic-resistant to anti-biotic-sensitive bacterial cells by cell-to-cell contact. This characteristic was explained by the discovery of DNA molecules called **plasmids**, which replicate independently of the cell's chromosome and are transferred during cell-to-cell contacts. Understanding the spread of antibiotic resistance was greatly facilitated by the earlier studies of bacterial conjugation (Figure 4.2). Indeed, the mechanisms governing plasmid transfer and conjugation proved to be related. The transfer of genetic information from one cell to another in conjugation turned out to depend on the presence of a plasmid in the donor cell.

Bacterial plasmids are circular, double-helical DNA molecules. They contain origins of DNA replication and genes that encode proteins that enable the plasmids to replicate (Figure 4.11). Some plasmids can replicate so efficiently that thousands of copies of the plasmid DNA are formed in each cell. Plasmids that confer resistance to one or several antibiotics do so because they contain genes that encode proteins that alter a cell's response to an antibiotic. Some plasmid genes encode enzymes that break down or otherwise modify particular antibiotic molecules. One example is an enzyme that destroys penicillin. Other resistance genes work in different ways. Tetracycline resistance, for example, results from a protein that interferes with the ability of cells to absorb tetracycline from their environment.

Plasmid DNAs can be readily isolated from cultures of bacterial cells in substantial (milligram) quantities and in pure form. Because of their ability to replicate in cells that harbor them, efforts were made to use them as vectors for introducing new DNA segments into cells. The joining of plasmid DNA to DNA inserts can be carried out in solution, in test tubes, much like the joining of inserts to phage vectors (Figure 4.9). In order to clone these inserts, however, it is necessary to introduce the recombinant plasmids into cells in which they are able to replicate (Figure 4.11). But, unlike phages, which enter cells from the surrounding fluid, plasmids are normally transferred directly from one cell to another. Methods were developed for promoting the uptake of plasmid DNA directly from the surrounding fluid into suitable bacterial host cells. With such methods in hand, individual cells could then be transformed by recombinant plasmid DNA and grown into colonies of cells containing novel, cloned molecules of recombinant DNA. Note that, because each host cell takes up only one plasmid, each colony has only one type of recombinant plasmid. Because many plasmid vectors are only a few thousand base pairs long, enrichment of eukaryotic DNA segments is even greater than with phage vectors. Plasmids were actually the first vectors used for molecular cloning in bacteria.

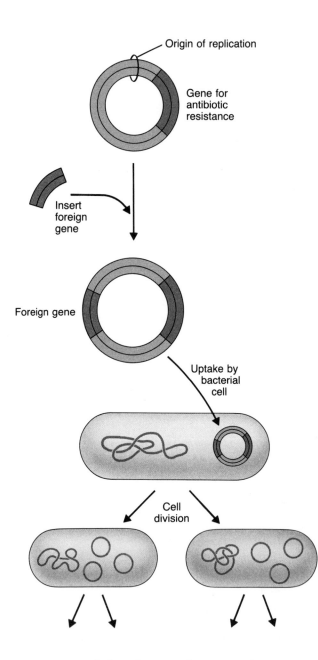

Figure 4.11   Bacterial plasmids are used as vectors for multiplying foreign DNA inserts.

Molecular Scissors

A N OFTEN unacknowledged reward from the advances made during the extraordinary period of biochemical and genetic discovery since 1950 was the identification of many enzymes that act on nucleic acids. Their isolation and characterization greatly facilitated the analysis of the structure and function of nucleic acids. Moreover, their availability made the recombinant DNA techniques possible. These include the enzymes DNA polymerase, RNA polymerase, reverse transcriptase, and DNA ligase mentioned in Chapter 2. But there are many more, including a whole catalog of nucleases—enzymes that catalyze the cleavage of the chemical bonds that hold DNA and RNA chains together. Some nucleases work only on DNA, others only on RNA, some on both types of nucleic acid. Some nucleases break nucleic acid chains in the middle. Others cleave off one nucleotide at a time, starting at one end or the other. Some cleave only double-helical chains, others only single strands (Figure 2.14). There are even enzymes that cleave an RNA strand only if it is in a DNA–RNA double helix; the DNA chain is left untouched.

Some of these enzymes, like the polymerases and ligases, are ubiquitous. The genes that encode these proteins, and the proteins themselves, occur in the cells of all organisms. Other enzymes are encoded and produced only in certain types of organisms, or only in phage-infected cells. All enzymes used for experimental purposes are extracted from cells and purified prior to their use. Our ability to understand the chemistry of nucleic acids and to manipulate pure DNA outside of cells depends entirely on the enzymes that cleave, modify, and join DNA molecules in specific ways. No purely chemical (that is, nonenzymatic) methods are known that can achieve the selectivity, the range, or the precision of enzymatic methods in the restructuring of DNA. It is interesting that most of these enzymes were discovered under circumstances unrelated to their uses in recombinant DNA experiments. Indeed, each of them has a vital role in the genetic chemistry of the organism from which it is derived.

Among the most extraordinary of the enzymes used for the manipulation of nucleic acids are the so-called **restriction nucleases**, whose discovery was one of the very important outcomes of genetic studies on bacteria and their phages. Each restriction nuclease recognizes and binds to a specific short sequence of base pairs in a DNA molecule and then cleaves both strands of the double helix at the binding site or a short distance from it.

The key observation leading to the discovery of restriction nucleases was made about 30 years ago. It was found that phages grown in one strain of bacteria often grow very poorly when they infect another strain of the same species. Moreover, the few phages formed during the inefficient infection of the second strain of bacteria can then propagate very well in this new strain, but they grow poorly in the original bacterial strain. This phenomenon, called restriction, was clearly correlated with some effect on the phages by the host cells. It was inferred that some phage component that is required for replication was modified by the host cells in a strain-specific manner. This modification was presumed to allow the phage to reproduce upon reinfection of the same host strain but to *restrict* it from growing in strains that do not contain the same modification system.

The modified phage component proved to be DNA. Genetic and biochemical analyses established that the modification involved the addition of methyl ($—CH_3$) groups to certain DNA bases and that the restriction resulted from cleavages of the unmodified DNA. Methylation occurs only within a short, specific nucleotide sequence (for example, at the first A in the sequence 5'-GAATTC-3'). "Restriction" is the result of cleavage by a nuclease between the G and the A, but only when the A is unmethylated. Thus, restriction of phage growth in an unfamiliar strain is caused by fragmentation of the unmethylated, invading phage DNA. If a small portion of the phage DNA gets methylated before degradation occurs, that methylated phage DNA can give rise to new phages. The low yield of progeny phages in such an infection is, therefore, a consequence of the very small amount of infecting DNA that (thanks to preemptive methylation) is spared from restriction cleavages. Because the progeny phage DNA is all properly methylated by the host cell's enzyme, it can be readily replicated when those phages infect the same *E. coli* strain again. However, because different strains use different target sequences for methylation and cleavage, growth of phages in one strain does not protect the phages against being cleaved in a second strain. A similar kind of methylation of the host cell's DNA protects it from degradation by its own restriction nuclease.

Modification (methylation) and restriction (cleavage) enzymes are encoded in the DNA of many different bacterial species, and even in phage and plasmid DNA. Invariably, modification and restriction systems are matched, because methylation and cleavage are directed at the same DNA sequence. In each bacterial strain or species, one or more different short nucleotide sequences are targets for methylation by specific enzymes. DNA that is appropriately methylated is not cleaved at that sequence by the matched restriction nuclease. Correspondingly, restriction nucleases cleave only those DNAs that are not methylated at their restriction sites.

Two features of restriction nucleases have proven to be extraordinarily influential in the analysis of the genetic and physical organization of complex genomes. First, different bacteria provide a great variety of enzymes that cleave DNA at different specific, short nucleotide sequences (Figure 4.12). Enzymes that recognize more than 100 different sequences are now commercially available as tools. DNA from any living organism can be cleaved, by these enzymes, into discrete sets of fragments, depending on the distribution of short nucleotide sequences that serve as target sites (Figure 4.13). The distinctive fragment pattern created by the action of each restriction nuclease on its special target sites in a DNA provides a unique fingerprint that characterizes the DNA.

In the laboratory, the kinds of DNA fragments produced by a restriction nuclease can be determined by separating them on the basis of length. This is accomplished readily by subjecting the fragments to an electric field. A solution containing the mixture of fragments is placed at one end of a thin gelatinous slab made either from an agarlike material or from polyacrylamide. An electric current is passed through the slab and, because they are negatively charged, the DNA fragments move through the slab toward the positively charged electrode. The shorter the DNA fragments, the faster they move. After the current is stopped, the DNA fragments are stained with a dye, which reveals the position of the fragments as bands in the gel. Because each restriction nuclease cleaves the DNA at a different set of sequences, the same large DNA molecule will yield a different assortment of fragment sizes with each enzyme.

The second important feature is that many restriction nucleases make staggered breaks on the two DNA strands within the restriction site (Figure 4.12)—that is, the cleavages are not exactly opposite each other but are separated by a few base pairs. Such cleavages create fragments with single-strand ends, which, because they will match with complementary single-strand ends, are referred to as "sticky" (or cohesive) ends. This fact makes it easy to join DNA fragments from different sources (Figure 4.14). Any two DNA segments that are obtained by cleaving DNA with the same restriction nuclease will have matching sticky ends. This will be so regardless of the source of the DNAs—for example, a vector DNA and a human DNA fragment. The two DNA segments can be mixed together and then joined by allowing the complementary single-strand ends to anneal. Then the strands can be connected through strong chemical bonds by the action of DNA ligase. This provides a relatively simple way to insert DNA fragments into phages, plasmids, or other potential vector DNAs.

100

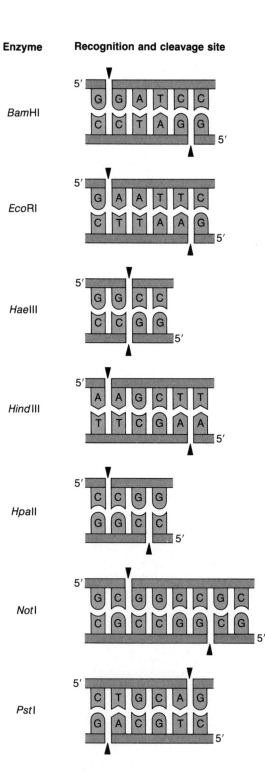

Figure 4.12 A group of restriction nucleases (named on the left) and the DNA sequences they recognize. The cleavage site on each strand of the double-helical DNA is marked with an arrow.

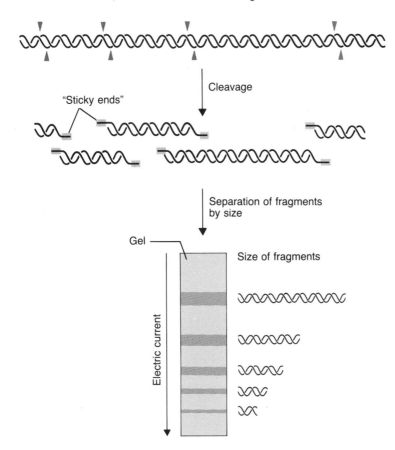

Restriction nuclease cleavage sites

"Sticky ends"

Cleavage

Separation of fragments by size

Gel

Size of fragments

Electric current

Figure 4.13 The cleavage by a restriction nuclease of a long DNA molecule containing several appropriate recognition sites (*arrows*) and the separation of the resulting DNA fragments in an electric field on a gel.

A scientist examines DNA fragment patterns

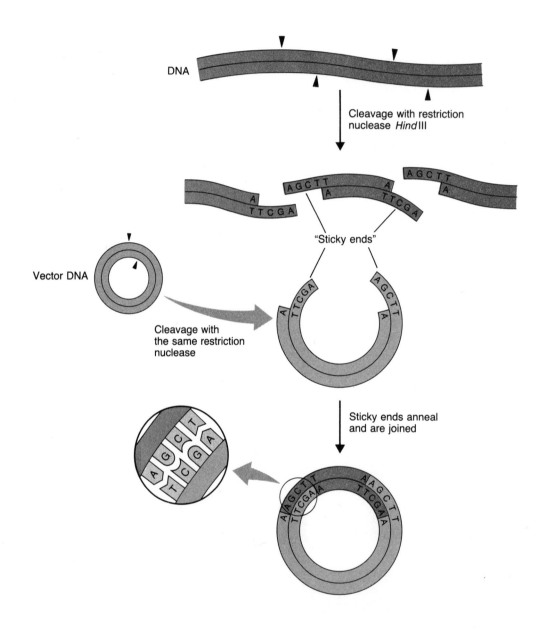

Figure 4.14   Many restriction nucleases make staggered cleavages in DNA; the resulting "sticky" single-strand ends (see also Figure 4.12) permit any two DNA segments to be joined.

Recombination Revisited

WHEN recombinant DNA techniques first came to public attention, many nonbiologists assumed that recombination was an artificial, unnatural process. This was one of the reasons for public skepticism about the advisability of recombinant DNA experiments. We know now that recombination is ubiquitous in nature.

Moreover, because of recombination, the organization of DNA sequences in genomes is not rigidly fixed; rather, recombination continuously promotes the formation of new arrangements. Most important, recombination is the process responsible for creating new genetic combinations in germ cells during sexual reproduction of eukaryotic organisms.

Since early in this century, recombination has also been the critical tool that has enabled us to understand the mechanisms of heredity. The centrality of recombination to genetic research will be apparent throughout the rest of this book.

CHAPTER 5

Cloning Genes

IN THE preceding chapter, we discussed the general principles of cloning and particularly the cloning of individual specific DNA molecules. But how are these principles actually put into practice? The first step in attempting to clone a particular gene from an organism's genome is to create a collection or library of recombinant DNAs that contains segments representative of the entire genome. Such libraries serve as the starting material from which a clone containing a single genomic fragment can be recovered.

Clones and Libraries

CONSIDER what steps are required to clone a gene. First, a large, complex genome is fragmented with a restriction nuclease. Then the entire population of different DNA fragments is mixed with a population of identical vector molecules, all of which have been cleaved at one position with the same nuclease. Recombinant DNA molecules are formed by annealing the complementary sticky ends on the vector molecules with those on the insert fragments (see Figure 4.14). The insert fragments are then sealed to the vector DNAs by ligation. These steps yield a population of DNA molecules, many of which contain the vector DNA with any one of a great number of genomic fragments. Other molecules in the population include vectors that lack inserts as well as some with multiple inserts. The final object is to clone the recombinant DNA molecule that contains the desired gene. An important question needs to be answered at this point: How many recombinant DNA molecules need to be recovered in order to ensure that every region of the genome being analyzed, including the gene in question, is represented in the collection?

Recall that there are 6 billion base pairs of DNA in one normal diploid human cell. That means that a minimum of 300,000 recombinant DNAs containing insert fragments with an average length of 20,000 base pairs would be needed to represent the entire genomic DNA. Cleavage of the DNA into fragments of larger or smaller size would require fewer or more recombinant DNAs to accomplish the same end. In practice, several times the theoretical minimal number of required recombinant DNAs are sought to assure full representation and to allow for the fact that many of the vector DNA molecules will not acquire an insert.

The next steps require separation of the individual members of the population of DNA molecules by cloning. Cloning involves several steps. First, the population of recombinant DNA molecules is mixed with cells in which the vector can replicate. Cells take up the DNAs; under appropriate conditions, each cell is likely to acquire only one molecule. Next, the suspension of cells is smeared thinly over the surface of a dish so that single cells are separated from one another. Meanwhile, the vector molecules—both those containing inserts and those without inserts—begin to replicate within the cells. If a phage or other viral vector is used, this process converts the assortment of vector molecules into a population of phage plaques on a plate of bacterial cells or of virus plaques on a plate of animal cells. When plasmid vectors are used, the individual DNAs are sorted into individual colonies of bacterial cells. In principle, all the recombinant DNA molecules are now cloned either in the form of a plaque or in the form of a colony. The next problem to be solved is to identify the desired clone and to expand it so that useful amounts of the recombinant DNA can be prepared. As a practical matter, it is advantageous to first get rid of all the plaques or colonies that contain vector molecules that lack inserts.

Most plasmid vectors include one or more genes encoding resistance to an antibiotic. These genes can be put to good use in distinguishing between a colony containing a recombinant DNA and one containing either an unaltered vector DNA molecule or no plasmid at all. A plasmid vector with two different genes for antibiotic resistance (for example, resistance to ampicillin and to tetracycline) is most useful in DNA experiments. Host cells containing such a plasmid grow in the presence of either or both antibiotics, but any cells that fail to acquire the plasmid do not. Thus, incorporating the antibiotics into the growth medium eliminates one whole group of unwanted cell colonies—those without plasmids.

The common plasmid vector illustrated in Figure 5.1 has such genes. The entire nucleotide sequence of the more than 4000 base pairs in this plasmid is known. The oval represents the site at which plasmid DNA

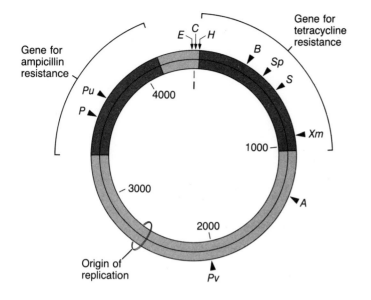

Figure 5.1   A circular bacterial plasmid DNA that serves as a vector.

replication begins. Other genes are not marked. The letters and arrows around the outside show sites that can be cleaved by a variety of restriction nucleases. Note that there is only one site for each nuclease. The vector DNA can be cleaved at any one of those sites by choosing the appropriate nuclease. Any one of the sites can be used for the insertion of a foreign DNA segment, so long as the vector and the insert fragments have matching sticky ends. The trick is to insert the foreign DNA segments within one of the genes for antibiotic resistance—for example, at *Pu* or *B*. Cell colonies with recombinant vectors can then be distinguished from those with un-altered plasmid vectors because cells containing the recombinants are no longer resistant to one of the antibiotics; the resistance gene has been disrupted by the insertion (Figure 5.2). A large variety of similar tricks are used with phage and viral vectors. Ultimately, one is left with only those clones containing recombinant DNAs—that is, those in which an insert is present in the vector DNA. Such a set of clones is called a **recombinant library**. Each member of the library, whether it is a bacterial cell, a phage, or a virus, contains a unique segment of the human (or mouse, or corn, or fly) genome. Such libraries can be stored so that they are permanent resources. Over time, many different genes can be isolated from the same library.

How can the desired recombinant DNA—the one carrying the inserted gene of interest—be identified within the library? Although a variety of tricks are routinely used to solve this problem, we describe only a few to

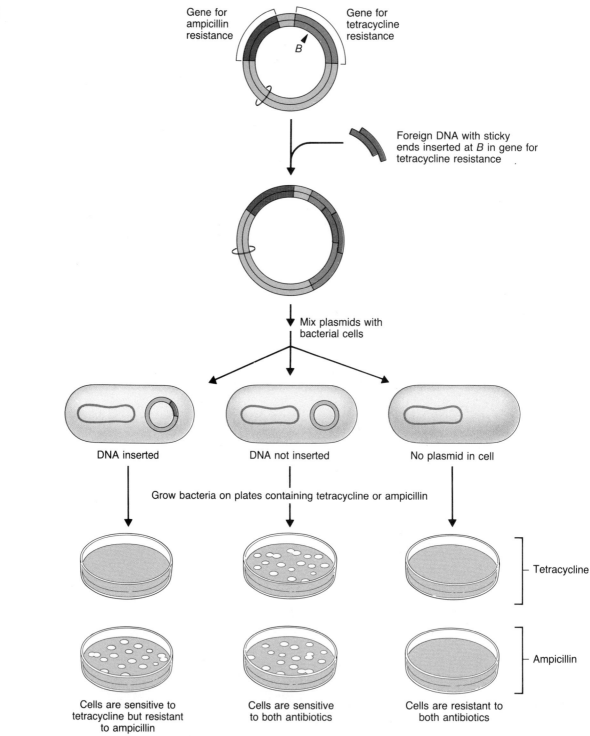

Figure 5.2 A plasmid vector with two genes for antibiotic resistance facilitates isolation of recombinants.

illustrate the general principles. The search for the single desired clone has to depend on a unique and detectable property of the insert. There are only two such classes of properties. One is the structure of the DNA insert—that is, the sequence of base pairs. The other is the protein encoded by a gene included in a cloned segment. These two unique properties are exploited in various ways to search through large numbers of clones and identify the one that is wanted. The methods are selective enough to allow one to find a single clone out of a few million, after a few days of laboratory work. Because each DNA segment and the protein it encodes is different, however, selection of a clone from a library depends on tailor-made procedures.

One of the most widely used techniques is also very simple. As with many selection procedures—indeed, as with many methods in molecular genetics—it depends on the unwinding and annealing of complementary DNA strands (Figure 2.9). Thousands of phage plaques or hundreds of bacterial colonies on an agar plate are transferred to a solid disk, usually of nitrocellulose, simply by laying the disk on the agar surface and lifting it up again (Figure 5.3). The positions of the plaques or colonies are undisturbed on the nitrocellulose imprint. The two strands of the DNA in the plaques or colonies are unwound, in place, by immersing the disk in an appropriate solution. Then the disk is placed in a solution containing a single strand of DNA (or RNA) whose sequence is complementary to one of the strands in the desired clone. The DNA (or RNA) in solution is tagged for easy detection, usually with a radioactive atom. Wherever the tagged single strand anneals to the proper recombinant, it yields a radioactive spot on the disk at a position corresponding to that of the recombinant on the original plate. The small amount of phage or bacterial colony that remains on the agar can readily be recovered. The tagged DNA or RNA in such an experimental setup is called a **probe**, because it is the means by which a nucleotide sequence can be found.

If the nucleotide sequence of the desired segment is unknown, and thus no probe is available, other identification procedures may be used. These depend on finding a clone that is synthesizing the protein corresponding to the gene of interest. This approach requires the use of special vectors that allow cloned genes to be transcribed and translated. Sometimes, the protein imparts a distinctive function to the host cells, and this serves as the basis for the detection. In other cases, the desired recombinant molecule can be identified because the protein being made can be detected with an antibody to the protein.

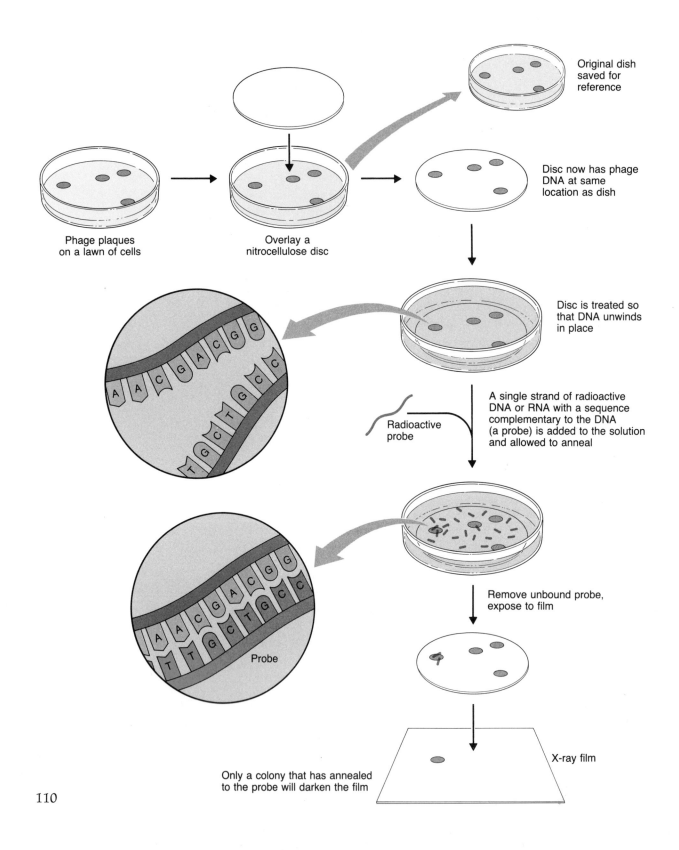

Original dish
saved for
reference

Disc now has phage
DNA at same
location as dish

Phage plaques
on a lawn of cells

Overlay a
nitrocellulose disc

Disc is treated so
that DNA unwinds
in place

Radioactive
probe

A single strand of radioactive
DNA or RNA with a sequence
complementary to the DNA
(a probe) is added to the solution
and allowed to anneal

Probe

Remove unbound probe,
expose to film

Only a colony that has annealed
to the probe will darken the film

X-ray film

110

Figure 5.3 (*facing page*)   Screening for a particular clone by annealing to a radioactive DNA or RNA probe.

Once the desired clone is identified, it can be picked off the agar plate with a needle. Cells carrying recombinant plasmid DNAs or plaques containing recombinant virus DNAs are then grown under appropriate laboratory conditions. Finally, the cloned recombinant DNA can be chemically purified from the cells or viruses. Once it is available in useful amounts, the cloned DNA containing the vector and gene can be used for a variety of laboratory manipulations. In this context, "useful amounts" means on the order of a milligram (a thousandth of a gram). This may not seem like much, but the sensitivity of modern methods is such that a milligram is more than enough for almost any imaginable experimental purposes. It is a simple matter to grow the cloned cells, or phages, or viruses again and again, providing an unlimited source of the recombinant DNA. There are simple and reliable methods for the permanent storage of libraries, colonies of cloned cells, and viruses. Only a single cell or virus particle is required to initiate growth and reproduction.

Cloning RNA

INDIVIDUAL types of RNA are difficult to study directly. More often than not, it is impossible to obtain a pure preparation of a single kind of mRNA. RNA is also quite unstable during laboratory manipulations. Moreover, no existing methods permit the facile, direct preparation of recombinant RNAs or the direct cloning of RNA molecules. The most convenient way to study RNA structure and function is to make a DNA copy of the RNA and clone the copy. This can be done in the laboratory using a reverse transcriptase. These enzymes assemble individual deoxynucleotides into DNA strands using RNA strands as templates. The mechanism is like that of other DNA polymerases except that the template is a single strand of RNA. Reverse transcriptases purified from retroviruses are available for routine laboratory use.

Reverse transcription yields a single DNA chain that is a copy of the RNA; the nucleotide sequence of the DNA is complementary to that of the RNA (Figure 5.4). Thus, such a DNA is called a **complementary DNA** or **cDNA**. A second DNA strand can be assembled by DNA polymerase, using the first cDNA strand as a template. The double-helical cDNA can then be cloned. Such cDNA clones are especially important for the study of mRNA structure. A large mixture of mRNAs—for example, all the mRNAs in a particular cell type, which may number in the thousands—

112

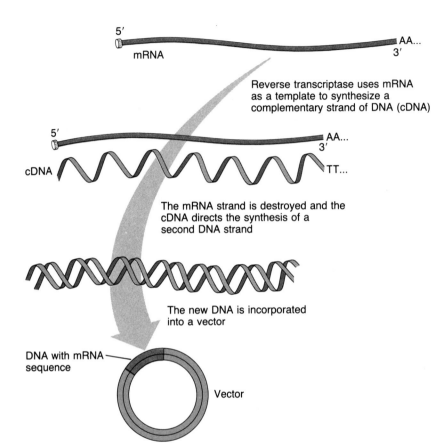

5′
mRNA
AA...
3′

Reverse transcriptase uses mRNA
as a template to synthesize a
complementary strand of DNA (cDNA)

5′
cDNA
AA...
3′
TT...

The mRNA strand is destroyed and the
cDNA directs the synthesis of a
second DNA strand

The new DNA is incorporated
into a vector

DNA with mRNA
sequence
Vector

Figure 5.4 Copying a messenger RNA with reverse transcriptase and inserting the copy DNA (cDNA) into a vector.

can be reverse-transcribed. The mixture of cDNAs can be recombined with vector molecules, and the resulting recombinant DNAs can be introduced into appropriate host cells and cloned. The process is virtually identical to that used with genomic DNA. The mixed population of recombinants constitutes a **cDNA library**. Individual cDNA clones can be isolated and characterized in much the same way as genomic DNA clones. Although determining RNA structures directly is tedious and difficult, the conversion of RNA into cloned cDNAs makes the analysis of RNA structure and function straightforward.

Why are cDNA libraries important? We have already pointed out that only certain genes are expressed in particular cell types or tissues or at particular times in the life of any organism. Moreover, genes and coding sequences represent only a small portion of the genomes of complex organisms, such as humans. On the other hand, the mRNAs present in a particular cell are derived from only a limited set of genes—those being actively transcribed. Regions of the genome that do not contain genes are not, for the most part, transcribed. Thus, a cDNA library represents expressed genes and is useful for analyzing which genes are expressed in a given cell type or tissue. As we shall see later, cloned cDNAs provided the first indications that the structure of mammalian genes was not what was expected.

Host—Vector Systems

CLONING of DNA segments, including genes and cDNAs, is most conveniently carried out in *E. coli* using plasmid and phage vectors that can multiply in the bacterial cells. Subsequent manipulations often require vectors that can deliver a cloned segment into eukaryotic cells. This is the case, for example, when the object of the research is to understand mechanisms of a gene's expression in a eukaryotic cell. A variety of vectors for use in yeast, plant, and animal cells have been devised. Frequently, they utilize DNA sequences from the genomes of viruses that infect eukaryotic cells. Each small extrachromosomal genome—bacteriophage, plasmid, and eukaryotic virus—is found in nature within a particular species and replicates only within cells of its natural host or within cells of closely related species. Therefore, the fundamental tool of molecular cloning is always a two-component system: a host cell and a compatible vector. In the process of developing suitable vectors for various experimental situations, many naturally occurring plasmid, phage, and viral genomes have themselves been extensively revised by recombinant DNA techniques. Two types of vectors are especially noteworthy.

Shuttle vectors are constructed so that they can replicate in cells from two different species. For example, yeast—*E. coli* shuttles contain two origins for DNA replication, one that works in the eukaryotic yeast cells and one that works in the prokaryotic cells of *E. coli* (Figure 5.5). They also contain one or more genes for antibiotic resistance. A yeast gene can be conveniently cloned—taking advantage of the ease of techniques for cloning in *E. coli*—and then the isolated recombinant DNA vector can be introduced into yeast cells to study expression of that gene.

The second very important class of vectors are the so-called **expression vectors**, which permit a cloned gene to be expressed—that is, to be transcribed and translated. Such expression vectors are very useful for screening

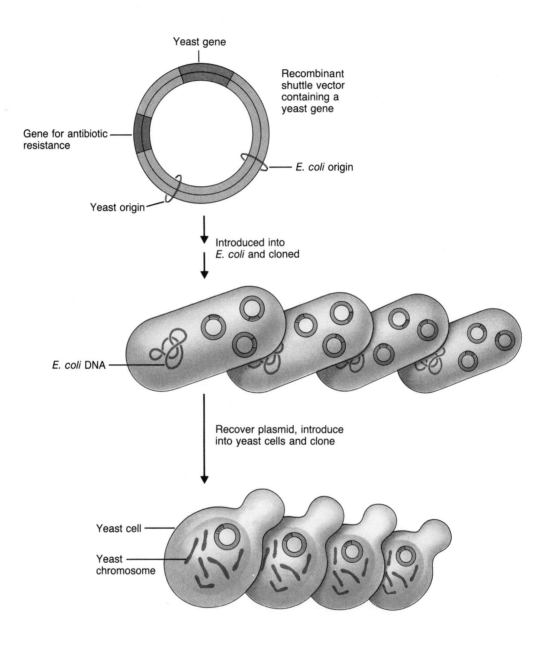

Figure 5.5 Shuttle vector for use in *E. coli* and yeast.

libraries when no appropriate RNA or DNA probe is available. Substantial amounts of the polypeptide encoded by a foreign gene can be produced in cells by expression vectors. This is critical if the aim of the cloning is to produce a polypeptide for use as a therapeutic agent or vaccine, or for use in research or manufacturing. The production of such proteins was one of the earliest goals of recombinant DNA research, and it is the basis for much of the thriving biotechnology industry.

Human growth hormone is a good example of an important therapeutic protein that was once difficult and expensive to obtain. The hormone had to be isolated from the pituitary glands of human cadavers. There was never enough hormone available to treat humans suffering from dwarfism. Once the gene for human growth hormone and its cDNA had been cloned, it became possible to introduce the coding sequences into a vector containing a promoter that is active in *E. coli* (Figure 5.6). In this case, the coding sequence for growth hormone was inserted into the vector close to the well-understood β-galactosidase promoter. After the recombinant vector DNA is introduced into *E. coli* cells, it replicates, yielding cells with a large number of vector molecules, each of which can express the gene for human growth hormone. The cells are allowed to grow and divide, forming a large population in which each cell is synthesizing human growth hormone. In an industrial vat, an *E. coli* culture amounting to thousands of liters can be grown, which, at roughly one billion cells per milliliter (equivalent to one trillion cells per liter), is thousands of trillions of *E. coli* cells, all making human growth hormone. Consequently, human growth hormone

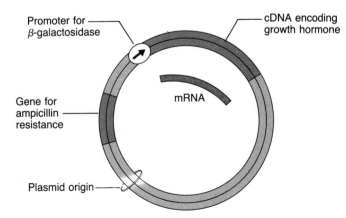

Figure 5.6 General features of an expression vector, using growth hormone coding sequences as an example.

is now relatively inexpensive and is available to all who need it. Similar strategies allow proteins from a variety of species to be synthesized abundantly in *E. coli*, in yeast, or even in animal cells growing under special laboratory conditions. These procedures have made available a large assortment of novel therapeutics: Tissue plasminogen activator for treatment of heart attacks; erythropoietin for treatment of anemia; interferons for treatment of cancer and hepatitis infections; human insulin for diabetes; and a variety of growth factors for stimulating blood cell formation. And there are many more in the works!

Mapping Cloned DNAs

THE FIRST step in the structural analysis of cloned DNA segments is to find out how the sites for cleavage by restriction nucleases line up on the DNA. As we have already seen, restriction nucleases provide a means to cut large DNA molecules into reproducible sets of relatively small DNA fragments. The fragments in the set can be separated from one another on the basis of size by electrophoresis through a semisolid gel (Figure 4.13). Generally, the shorter a double-helical fragment, the faster it moves in the electric field. Fragments from 20 to millions of base pairs in length can be separated, depending on the way the electrophoresis is carried out. Thus, given some marker fragments of known size, one can readily determine the size of a new fragment with a ruler and a simple graph. Less than a microgram (a millionth of a gram) of DNA is usually sufficient for such analyses, because the DNA fragments are readily detected when stained with an appropriate dye.

The set of fragments produced with a particular restriction nuclease provides a characteristic fingerprint of the initial DNA. By analyzing the fragments produced by several restriction nucleases alone and in combinations, the order of the fragments within the original DNA molecule can frequently be deduced. The result is a physical map of the DNA, which, if sufficiently detailed, uniquely defines the original DNA molecule (see the map of the sites at which restriction nucleases can cleave the plasmid in Figure 5.1). Recombinant vectors are generally reasonably small, so the job of constructing the physical map is not too difficult. However, large DNA molecules treated with restriction nucleases yield complex sets of fragments. Specially designed computer programs have proven to be useful for creating physical maps from analyses of such sets of fragments.

Often, the second step in the structural analysis is to determine where the gene lies within the cloned DNA fragment—for example, within which restriction nuclease fragment—and whether all or only part of the desired gene is present. Here again, the basic tool is annealing of the unwound DNA fragment with cDNA (or RNA) probes. Electrophoretic gels, like those shown in Figure 4.13, can be blotted directly onto a solid sheet of inert material without disturbing the distribution of separated DNA fragments (Figure 5.7). After the gel is placed in contact with a sheet—of nitrocellulose, for example—an appropriate solution is drawn through the gel and the nitrocellulose, carrying the DNA into the nitrocellulose sheet, where it is physically trapped. The two strands of the DNA are unwound on the sheet and, just as in the method that is used for finding a clone (Figure 5.3), the sheet is immersed in a solution containing a tagged probe of single-stranded DNA (or RNA) complementary to the sought-after segment. If, for example, the probe is tagged with a radioisotope, a radioactive band on the sheet reveals the position, and hence the size, of any DNA fragment that contains the gene of interest. This procedure allows for positioning the segment of interest on the map.

The procedure known as **DNA blotting**, which is described in Figure 5.7, is used widely, and not only to characterize cloned DNAs. One of the most important applications of the method is its use in identifying a DNA fragment containing a single gene of interest among many, many other DNA fragments. A mixture of fragments obtained after treatment of total cell DNA with a restriction nuclease can be separated according to size by gel electrophoresis. If the genome is very large, like the human genome (6 billion base pairs), millions of fragments of various sizes are produced. After staining the gel with a dye, the individual DNA fragments are obscured by a smear of DNA extending from the largest to the smallest possible fragments (Figure 5.8). This is because no one size of fragment (of the millions of fragments produced) is abundant enough to be seen as a discrete band. But if such a gel is blotted to a nitrocellulose sheet and incubated with a radioactive DNA or RNA probe, the corresponding genomic DNA fragment will appear as a single radioactive band on the strip at a position corresponding to its size (Figure 5.8). Only a few millionths of a gram of total genomic DNA suffices. This technique, **genomic blotting**, is critical to modern genetic mapping. It is also the basis of diagnostic techniques for many genetic and viral diseases, and it is used in forensic medicine to identify both victims and criminals. In later chapters, we will refer to blotting again and to very similar procedures that are used for the detection and identification of RNA and proteins.

118

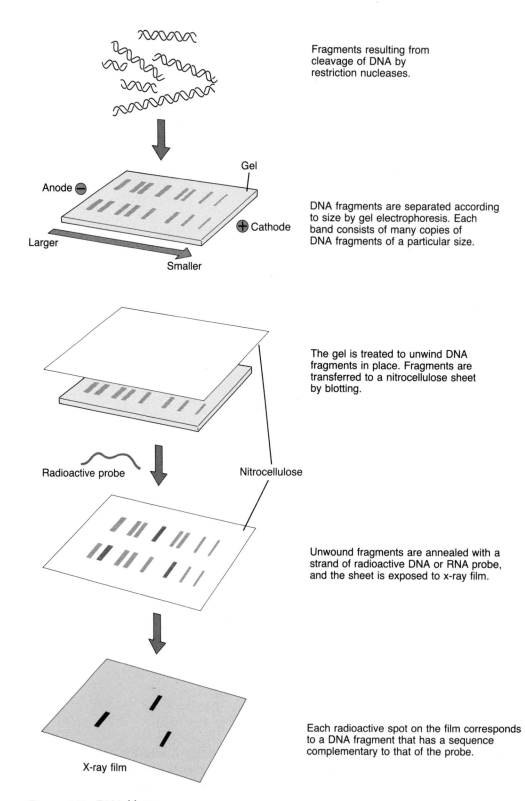

Fragments resulting from cleavage of DNA by restriction nucleases.

DNA fragments are separated according to size by gel electrophoresis. Each band consists of many copies of DNA fragments of a particular size.

The gel is treated to unwind DNA fragments in place. Fragments are transferred to a nitrocellulose sheet by blotting.

Unwound fragments are annealed with a strand of radioactive DNA or RNA probe, and the sheet is exposed to x-ray film.

Each radioactive spot on the film corresponds to a DNA fragment that has a sequence complementary to that of the probe.

Figure 5.7 DNA blotting.

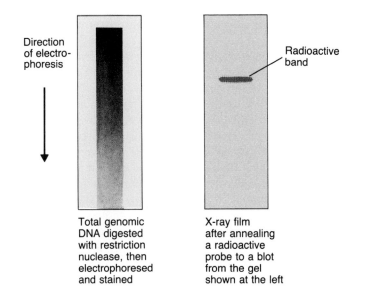

Direction of electrophoresis

Radioactive band

Total genomic DNA digested with restriction nuclease, then electrophoresed and stained

X-ray film after annealing a radioactive probe to a blot from the gel shown at the left

Figure 5.8 Blotting genomic DNA.

Fine Anatomy of a Cloned DNA

ULTIMATELY, understanding the structure, function, and evolutionary history of a particular segment of genomic DNA depends on knowing its nucleotide sequence, which can be determined from its cloned counterpart. Fast, accurate methods for determining DNA sequences were invented in the late 1970s, soon after recombinant DNA techniques were developed. In principle, there is now no limit to the length of DNA that can be sequenced. The sequences of DNA segments many hundreds—even thousands—of nucleotides long are determined routinely. In the hands of an able person, a DNA sequence several thousand base pairs long can be determined in a week's time and with an accuracy of more than 99.9 percent (one misidentified base in a thousand). Furthermore, automated instruments capable of producing sequence information are now available. They are expected to increase the efficiency of determining DNA sequences markedly and to relieve the tedium of this kind of work. However, even with present methods, about 89 million base pairs of DNA sequence from many different organisms have already been determined (as of May 1992). The data are stored in accessible, centralized computer banks in the United States and in Europe. Of the 89 million base pairs, about 16 million represent scattered segments of the human genome. This is a start on the big plan to determine the sequence of the entire human genome in coming years.

Biologists' yearnings for chemical precision are nourished by the accumulating sequence information. These data are useless, however, unless the biologically significant features of the sequences can be recognized. Some help is available from computers: There are programs that readily identify extended regions of translatable codons that might hint at the presence of a protein-coding gene. The programs can also find regulatory sequences, such as promoters and terminators. The search programs are especially effective in identifying repetitive DNA sequences that are common in eukaryotic genomes, similarities between sequences, and recognition sites for restriction nucleases. Analogous programs can compare a sequence with all the other sequences in the data bank. There are programs that reveal how a DNA sequence could be translated into a polypeptide sequence, or that indicate the many different DNA sequences that could, in theory, encode a given polypeptide (remember that the genetic code has more than one codon for most amino acids). Other programs predict the regions in a single RNA strand that are most likely to form double-strand regions through complementary base-pairing, thereby providing models for such folded molecules as tRNA or rRNA. As the amount of sequence data expands, the need for increasingly sophisticated computer technology is stimulating collaboration between mathematicians, computer scientists, and biologists.

Reverse Genetics

EXPRESSION of a gene requires that it be transcribed into RNA. To study whether a cloned gene is expressed in particular cells, the mixture of RNAs present in those cells is prepared and separated, by gel electrophoresis, into individual RNAs according to size. The gel is blotted and the blot is analyzed with a radioactive DNA probe containing segments of—or all of—the cloned gene. If a radioactive band is seen, it means that an RNA complementary to that gene is present. If the RNA is of an appropriate length (genes vary a good deal in length), there is a good chance that it is the sought-after mRNA. To be certain that it is indeed the desired mRNA, other procedures can be used to identify the protein encoded by that RNA.

The biological activity of a cloned gene can sometimes be studied by strictly chemical methods. Using RNA polymerase and crude mixtures of ribosomes, tRNAs, and other essential ingredients obtained from broken cells, the cloned gene can be transcribed into RNA and translated into protein in the test tube. Clearly, such analyses require that the cloned gene include a promoter sequence. Note that the whole flow of genetic information from DNA to RNA to protein can thus occur outside of living cells.

Recombinant DNA techniques can be used to replace or alter the sequence of the regulatory segments that control the expression of a particular cloned gene. Conversely, one regulatory segment can be joined to different genes (as is done in the construction of expression vectors). The biological functions of such constructions can be tested in cells as well as in test tubes. Such tests can be carried out in animal and plant cells growing in laboratory dishes because these cells can take up recombinant DNA molecules from the surrounding medium just as bacteria can, although less efficiently. The uptake of DNA by eukaryotic cells can be made more efficient if the cells are subjected to an electric shock or if the DNA is injected into the nuclei of the test cells.

Most often, the cells will express a cloned gene present in the introduced DNA, so long as the regulatory elements are functional in the cell. Bacterial regulatory sequences are required for expression in bacteria, even when the coding sequences are derived from a eukaryote. If the recombinant DNA contains a eukaryotic promoter at the proper location, both prokaryotic and eukaryotic coding sequences can be expressed in eukaryotic cells. Finally, it is possible to make specific changes in the nucleotide sequences of cloned DNAs—either in coding regions or in regulatory sequences— and to test the function of such modified structures in or outside of cells. Genes with altered DNA sequences—new alleles—can be designed and prepared at will. Chapter 12 describes how such purposefully modified genes can also be introduced into whole plants and animals. Thus, we are no longer dependent on the chance isolation of mutations, as was the case in classical genetic analysis.

CHAPTER 6

Genes Are More Interesting
Than We Thought

MOLECULAR cloning has made virtually every segment of every genome accessible. Consequently, the impasse previously posed by the complexity of eukaryote genomes and the difficulty in their genetic analysis has been overcome. Now, specific pure DNA segments from any genome can be obtained in sufficient quantities to allow complete chemical description. The use of restriction nucleases and the analysis of nucleotide sequences make such descriptions almost routine. A technique called **genome walking** identifies individual cloned DNA segments that overlap one another in a genome. While the size and complexity of eukaryotic genomes remain formidable (Table 3.1, p. 76), there is no longer a substantive barrier to knowing the molecular details of their organization and structure. The remaining obstacles are the realities of time and resources. It took about five person-years to determine the sequence (published in 1985) of the 50,000 base pairs in the genome of bacteriophage lambda. Major improvements in methods of nucleotide sequencing are now being made. The nucleotide sequence of the 315,000 base pairs that make up the entire yeast chromosome III was reported in 1992. The 4 million base pairs of the *E. coli* chromosome are within reach, and a small human chromosome is only about ten times larger. Some people think that we will know the entire sequence of the human genome early in the next century, but this prediction relies on major technological innovation and a highly concentrated effort.

There are many reasons for cloning genes and other DNA segments and determining their sequences. One is to elucidate the function of their genomic counterparts in the cell. The analysis of cloned sequences can often reveal the presence of specific regulatory elements. The presence of nucleotide sequences that lack stop codons (that is, that have open reading frames) suggests that the cloned segment encodes a protein. But the nucleotide sequence of a gene and its flanking segments do not tell us how the gene works or how its expression is regulated during the development of a complex organism, in the specialization of different cells, and in response to environmental changes. Neither does the sequence reveal how the expression of genes is coordinated to ensure the intricate physiological balance characteristic of healthy cells and organisms. Nor can we deduce how the production of a faulty protein by a mutated gene, or the altered expression of a gene with a mutated regulatory signal, disrupts normal function or produces disease. To comprehend the physiological significance of the molecular detail requires biological and biochemical analysis. Here, too, recombinant DNA techniques provide a powerful experimental approach in the types of experiments referred to as "reverse genetics" at the end of Chapter 5. Because these concepts are so important to everything that follows, we will review them here.

Cloned genes can be introduced into cells by a variety of special techniques. Often, such newly introduced genes are expressed and even respond to the same regulatory signals that influence that gene's function normally—for example, the way in which the β-galactosidase gene in *E. coli* responds to glucose and lactose (p. 73), or the way in which some mammalian genes respond to estrogen, or some plant genes respond to light. Thus, the behavior of mutant and normal genes and regulatory signals can be assessed after their introduction into cultured cells. Both naturally occurring mutant alleles and experimentally constructed mutants can be compared. Thus, a very special consequence of the emergence of recombinant DNA technology is that the interdependence of a gene's structure and its activity can be analyzed for virtually any gene. It is now beside the point that complex organisms are not always amenable to the types of genetic and biochemical analysis that proved to be so useful for *E. coli*. Indeed, the new techniques allow even deeper understanding than the techniques previously used for bacteria, and for that reason they are being applied now to bacteria as well as to eukaryotes. Thus, the new techniques expand the potential of genetic research in extraordinary ways.

The application of classical genetic and chromosome analysis to organisms like humans was necessarily limited. Now such genomes can be studied at the fundamental level. A few micrograms of DNA isolated from mammalian white blood cells and converted into a gene library can be used to determine the relation between the structures of a large number of different genes and their contribution to cellular physiology. Moreover, the universality of recombinant DNA techniques means that the genomes of all obtainable contemporary organisms (and even ancient mummies, or extinct organisms whose remains can yield DNA) are accessible, and their similarities and differences can be compared. As we shall see later, comparisons of gene structure among different species provide important clues about the history of genome evolution and add a new dimension to the classical tools of the evolutionary biologist.

These grand and sweeping statements are not just wishful thinking. Admittedly, the promise of recombinant DNA techniques is only now beginning to be realized, and many—perhaps most—features of eukaryotic genetic systems remain to be discovered. But even in the short span of almost 20 years since the methods were developed, several truly startling aspects of eukaryotic genomes have been discovered.

Our earliest notions about eukaryotic gene structure were predicated on prokaryotic systems. Some biologists were so bold as to predict that what was true for *E. coli* genes would be equally true of elephant and human genes. In one sense, that is true, insofar as the general mechanisms of transcription, translation, and the genetic code, for example, are common. However, detailed analyses of many genes and genomes has revealed profound and remarkable differences between the two types of genetic systems. Even though eukaryotic and prokaryotic genes share certain structural and regulatory features, eukaryotic genes possess a complexity that drastically influences how they are expressed and regulated.

Overall, eukaryotic gene expression is far from being fully understood. When the whole story is known, it is certain to include aspects that have not yet been imagined. Even from the present perspective, it is clear that evolution has yielded a wide variety of mechanisms to control genomic activity. The remainder of this chapter highlights some of the major insights that have emerged from the application of recombinant DNA methodology to eukaryotic genes.

Gene Structure

LONG before eukaryotic gene structure could be studied directly, it was known that the genetic systems of prokaryotes and eukaryotes share certain fundamental properties. Their genetic material is DNA, the DNA is replicated in similar ways, genetic information passes from DNA to RNA to protein, and the genetic code is the same. It was expected—correctly, as we now know—that eukaryotic organisms, especially multicellular ones, would have more genes than prokaryotes. And it was widely assumed—

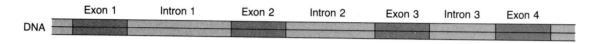

Figure 6.1 The organization of exon (coding) and intron (noncoding) sequences in a eukaryotic gene.

incorrectly—that the basic structure of the genes and the organization of genetic information in the two types of genomes would be similar. These two assumptions were demolished almost as soon as the first eukaryotic genes and genomes were analyzed.

By the late 1970s, it was clear that the protein-coding sequences in a eukaryotic gene do not necessarily consist of a single contiguous stretch of DNA, as they do in a bacterial gene. Instead, the coding region is often discontinuous, being interrupted by stretches of noncoding DNA; such noncoding DNA segments are called **intervening sequences** or **introns**, and the coding segments of genes—those that generally direct polypeptide synthesis—are referred to as **exons** (Figure 6.1).

The first inklings of the existence of introns emerged from studies of animal viruses. Surprisingly, the nucleotide sequences of various viral mRNAs did not line up with the corresponding nucleotide sequences in the virus's DNA. Indeed, the mRNAs seemed to be composed of nucleotide sequences present in discontinuous blocks in the viral DNA from which they were transcribed. Similar findings were made in mammalian cellular genes. For example, the mRNA encoding the β-polypeptide chain of hemoglobin (the same chain that is mutated in sickle-cell disease) annealed to several DNA fragments in a restriction-nuclease digest of total cellular DNA. Together, the length of these fragments added up to many more base pairs than the number of nucleotides in the mRNA. This too suggested that the DNA segments transcribed into the β-polypeptide mRNA were not contiguous as they were in all the prokaryotic genes known at that time. Instead, it appeared that a gene's coding sequence was dispersed over several separate regions of the chromosomal DNA.

Now we know that most (though not all) of the genes that code for polypeptides in vertebrates and plants are interrupted by at least one intron, and some by very many more (Table 6.1). Introns occur less frequently in the genes of yeast and invertebrates. Genes that encode functional RNA

Table 6.1 *Introns in a Few Representative Genes**

Gene	Organism	Exons Total bp	Introns Number	Introns Total bp
Tyrosine transfer RNA	Yeast	76	1	14
Uricase subunit	Soybean	300	7	4,500
β chain of hemoglobin	Mouse	432	2	762
Dihydrofolate reductase	Mouse	568	5	31,500
Erythropoietin	Human	582	4	1,562
Zein	Corn	700	0	0
Phaseolin	Bean	1,263	5	515
Hypoxanthine phosphoribosyl-transferase	Mouse	1,307	8	32,000
Adenosine deaminase	Human	1,500	11	30,000
Cytochrome *b*	Yeast (mitochondria)	2,200	6	5,100
Low-density lipoprotein receptor	Human	5,100	17	40,000
Vitellogenin	Toad	6,300	33	20,000
Thyroglobulin	Human	8,500	>40	100,000
Clotting factor VIII	Human	9,000	25	177,000
Fibroin (silk)	Silkworm	18,000	1	970

* The genes are named according to the protein they encode. The genes were chosen to illustrate the diversity of gene structure and the large amount of DNA devoted to introns.

molecules (like tRNA or rRNA) may also have introns, but interruptions are less common in these genes than they are in genes that encode mRNAs. In many genes, the amount of DNA in the introns exceeds the amount in the exons by as much as a factor of ten, and, in some cases, the amount of DNA in a gene's introns may be 100 or more times greater than that in its exons.

It is helpful to remember an important difference between prokaryotes and eukaryotes. In prokaryotes, there is no physical barrier between the site of DNA transcription and the ribosomes, which are necessary for translation. Ribosomes bind to mRNA and begin protein synthesis even before transcription is completed. In eukaryotes, however, mRNA (as well as tRNA and rRNA) is formed in the nucleus, where there are no ribosomes. In order to be translated, therefore, an mRNA must first be transported across the nuclear membrane into the surrounding cytoplasm, the site of protein synthesis. Early work on the RNA within the nuclei of eukaryotic cells produced some puzzling observations. The RNA molecules were, on the average, much longer than expected and were very heterogeneous in size. Moreover, most of the RNA made in the nucleus was not transported out to the cytoplasm.

The existence of introns and the enormous variation in their number and size explained these oddities. When the nuclear genes are transcribed, RNA polymerase begins to copy them at the position corresponding to the start (the 5' end) of the mRNA. The enzyme transcribes through the gene's exons and introns until it reaches the end of the gene (Figure 6.2). Because the RNA strands made in the nucleus contain both exons and introns, and because their sizes vary so much, it is not surprising that the RNAs are so heterogeneous in size. In contrast, mature mRNAs in the cytoplasm are completely free of introns. Special mechanisms exist to ensure that only intronless mRNAs leave the nucleus for the cytoplasm.

Introns in RNA transcripts made in the nucleus are removed by a mechanism called **splicing**. In the process, the RNA is cut at the junction between each exon and intron, and the two flanking exons are joined (spliced) to generate a continuous stretch of exons that contains the coding sequence (Figure 6.2). Although this process is simple in design, an elaborate cellular machinery is needed to ensure the fidelity of splicing. Only if the cleavages at the two ends of an intron are precise, and only if all the exons are joined to each other intact and in the correct order, can the mRNA be translated into its protein product. Not a single nucleotide in any exon can be lost.

Small nuclear particles—"snurps," as they are called—catalyze the splicing process in most instances. Snurps are composed of proteins and special small nuclear RNAs (see Table 2.1 on p. 50). They recognize and bind to characteristic short nucleotide sequences that occur within and at the two ends of all introns. After binding at the intron-exon junctions, the snurps cleave the RNA chain and then join the appropriate cut ends to form a continuous RNA chain. Remarkably, however, some introns can splice themselves out of their RNAs without the help of snurps or any protein. The intron folds itself up in a special way, so that cutting at the intron boundaries and joining of the exons occur spontaneously, without

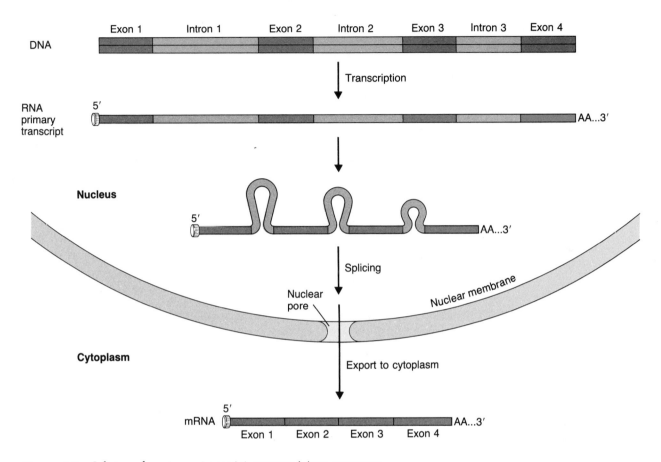

Figure 6.2 Splicing of a primary transcript removes intron sequences and joins the exons to yield a messenger RNA.

outside intervention (for example, by proteins). Such self-splicing is generally believed to be the evolutionary forerunner of present-day splicing mediated by snurps.

Most often, splicing results in the excision of all the introns and the conservation of all the exons in their original order as a continuous sequence in a single mRNA. However, splicing may occur in more than one way: All or part of an exon can be treated as an intron—that is, it can be removed during the splicing process. Consequently, several different mRNAs can be formed from the same RNA chain (Figure 6.3). In this example, two mRNAs are formed depending upon which exons are joined

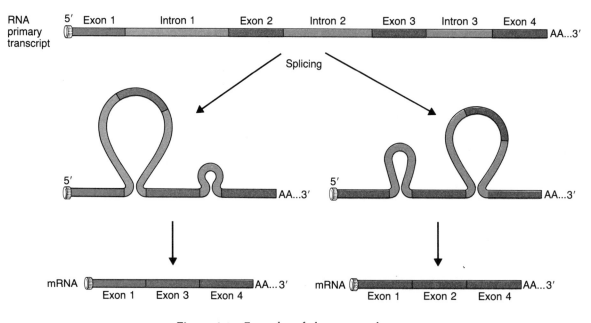

Figure 6.3 Examples of alternative splicing.

together. Each mRNA has a different subset of exons and, therefore, each encodes a protein with a different amino acid sequence. Such alternative splicing is of considerable importance because it provides an efficient way of producing diverse proteins from a single gene without altering the gene's DNA structure. Moreover, alternative splicing can be regulated, so that a gene can be expressed one way at one time during development or in one kind of tissue, and in another way at a different developmental stage or in a different tissue.

The difference between the nucleotide sequence of a eukaryotic gene and its corresponding mRNA has some important implications. A cDNA clone prepared from bacterial RNA yields the structure of the corresponding mRNA and, by inference, the structure of the gene. In contrast, a cDNA clone representing a eukaryotic mRNA does not give the structure of the corresponding gene because the mRNA lacks the introns that are present in the genome. However, comparisons between the structures of cDNAs and their corresponding genes give important information about the position and length of the introns. This is the way the data in Table 6.1 were acquired.

Sophisticated Switches

THE ANALYSIS of eukaryotic gene structure and function revealed that the machinery and signals that govern transcription are considerably more complex than bacterial regulatory signals. Prokaryotes contain a single kind of RNA polymerase, which transcribes all types of genes, and three different kinds of regulatory elements are used to regulate the transcription of their genes. First, there are the promoter sequences that determine where and how efficiently transcription begins. In *E. coli*, for example, promoters are defined by two short stretches of DNA that occur about 10 and 35 base pairs before the site at which transcription begins. A second kind of element defines the end of a gene (or group of genes) and triggers termination of transcription. Finally, there are DNA sequences around the promoter that are recognized by specific regulatory proteins, such as repressors and activators, to modulate transcription. Examples of these were described in connection with the *E. coli* β-galactosidase gene on page 71. All these regulatory sequences depend upon interactions with proteins to influence the expression of neighboring coding sequences.

Eukaryotes differ from prokaryotes in having three different RNA polymerases. Each RNA polymerase is responsible for transcribing a distinct class of genes. Moreover, each class of genes is associated with different and specific transcriptional control and terminator signals, and these can occur in a variety of locations, and even at various distances and directions from the sites where transcription starts and stops.

Genes for rRNA are transcribed by RNA polymerase I. The enzyme acts in conjunction with accessory regulatory proteins that recognize DNA sequences on the 5′ side of genes encoding rRNA. Some of these proteins function in a species-specific manner. For example, the rRNA genes of fruit flies are not transcribed by the RNA polymerase I system of human cells. RNA polymerase II transcribes all genes encoding proteins and most genes encoding the small nuclear RNAs in snurps. RNA polymerase III transcribes genes encoding tRNAs and a small rRNA. These enzymes also depend on a variety of additional regulatory proteins for their function. In contrast with the species specificity of RNA polymerase I, RNA polymerases II and III transcribe appropriate genes from any species, regardless of their origin.

Much of the ability of an RNA polymerase to transcribe the correct gene at the right time stems from the ability of different kinds of regulatory proteins to bind to one or several different short DNA sequences in the vicinity of the gene. The number and location of the bound auxiliary proteins influences the efficiency with which an RNA polymerase begins to transcribe the gene. In this sense, the basic logic of transcription regulation is similar in prokaryotes and eukaryotes: It involves the precise recognition of specific DNA sequences by special sets of binding proteins. A great deal of research is being done on these mechanisms, and much remains to be done. Already, there are hints that certain of these sequence-specific binding proteins are the key elements for directing the orderly process of embryonic development.

The detailed mechanism whereby the transcription of eukaryotic genes is terminated is presently unclear. Here too, however, depending on which RNA polymerase is used, different characteristic nucleotide sequences serve as termination signals. These sequences too are targets for specific proteins that mediate the termination event.

Messenger RNA (mRNA), tRNA, and rRNA participate in protein synthesis in the cytoplasm. Thus, almost all the RNA molecules transcribed in eukaryotic cells must be transported across the nuclear membrane and into the cytoplasm before they can fulfill their proper roles. A few small RNAs remain in the nucleus, where they are required for splicing, and for the modification of rRNAs and tRNAs to their functional forms.

Several steps are required to convert the RNA transcribed from a protein-coding gene into an mRNA (Figure 6.4). Besides being spliced to remove introns, eukaryotic mRNAs always have a special addition, referred to as a "cap," at the 5' end. Caps increase the efficiency of translation by allowing ribosomes to bind to the mRNA. The third modification associated with mRNA formation occurs at the 3' end of the RNA chain. Transcription by RNA polymerase II generally proceeds well beyond the 3' end of a protein's coding sequence. Instead of termination, the 3' end of the mRNA is formed by cutting the RNA chain about 20 nucleotides beyond a specific sequence (AAUAAA) that follows the end of the coding region. After the cleavage, the 3' end of the RNA is modified by the addition of 50 to 200 adenines to form a poly-A tail. Thus, the mature form of mRNA carries a long tail that is not present in the gene.

The expression of some genes is affected by changes in DNA structure that do not alter the normal nucleotide sequence. One such change alters the cytosine bases in -C-G- sequences near the 5' end of the gene. The modification consists of the addition of a methyl ($-CH_3$) group to the cytosine base, a reaction that is carried out by special methylating enzymes. Increased methylation in the region at the 5' side of a gene often correlates with reduced or absent gene expression, while low levels of methylation in that region correlates with high levels of expression. Gene expression is also influenced by the structure of the protein-bound form of DNA—

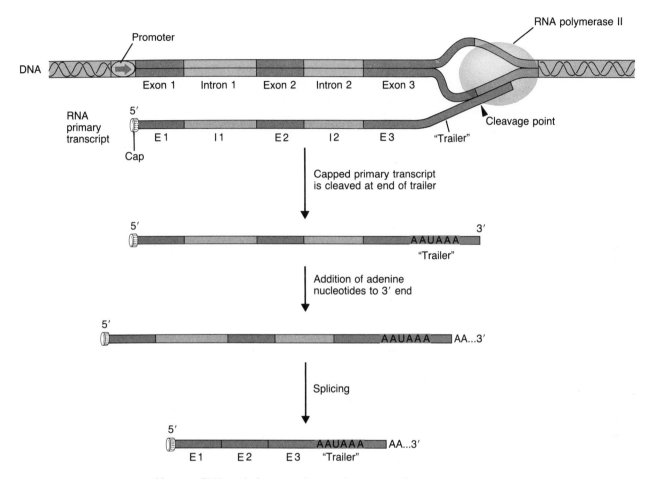

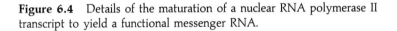

Figure 6.4 Details of the maturation of a nuclear RNA polymerase II transcript to yield a functional messenger RNA.

chromatin. Although the change itself is still poorly understood in chemical terms, the genes can become inaccessible to the transcription machinery, because chromatin can be folded into highly compact structures. By contrast, unfolding the compact form of chromatin frequently activates transcription in that region.

Numerous studies with different genes indicate that transcriptional on–off switches are the dominant means for controlling gene expression in

133

prokaryotes. In special circumstances, however, other mechanisms also come into play. They include, for example, control of transcription termination, control of translation, and regulation of the length of time mRNAs remain functional before they are destroyed. On–off switches are also critical in regulating the transcription of eukaryotic genes. At present, there is only suggestive evidence that transcription termination is also regulated in eukaryotes. But there are many indications that the rate of protein production can be governed by regulating the rates of mRNA translation and of mRNA breakdown. However, special opportunities for regulating the flow of genetic information arise from the unique features of eukaryotic cells and the structure of their genes and genomes. For example, the initial RNA transcript is unable to function unless its exon sequences are properly spliced together. Similarly, transcripts destined to become mRNAs must be modified at both ends with caps and poly-A tails for them to function in protein synthesis. Messenger RNAs also must cross the nuclear membrane into the cytoplasm before they can be translated. Each of these events is a point at which regulation may occur.

What Is a Gene?

AS WE come to understand more and more of the structural complexity of genes, the original definition of a gene becomes less and less applicable. Certainly, a contemporary definition of a gene must go beyond saying that it is a unit of heredity governing a particular trait. However, arriving at a single, consistent, molecular definition is not simple. For example, does an intron count as part of a gene? In some instances, a sequence can be an exon for one of the gene's proteins and an intron when alternative splicing leads to another protein. Is a regulatory sequence a part of a gene? Some regulatory sequences are thousands of base pairs away from the genes they control. What about a noncoding sequence that flanks a gene's coding region, gets transcribed, and becomes part of the mRNA— is it part of the gene? In fact, several different possible definitions of the term gene are plausible, but no single one is entirely satisfactory or appropriate in every case. At best, it is only possible to describe the elements that usually occur in a functional unit of heredity. Before doing that, a few of the features of genes should be reviewed.

Most genes include DNA nucleotide sequences that will be translated into proteins from the corresponding mRNA by ribosomes and tRNA,

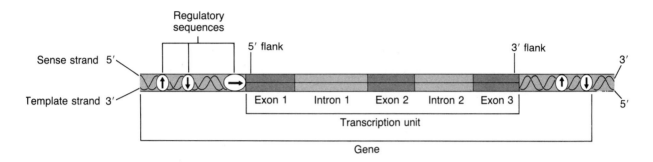

Figure 6.5 The DNA segments that constitute a gene.

according to the genetic code. Other genes are transcribed into RNA molecules that do not encode proteins but that function as part of the machinery for making proteins.

With these notions in mind, we might define a gene as a combination of DNA segments that together constitute an expressible unit—that is, a unit whose expression leads to the formation of either a functional RNA or a polypeptide (Figure 6.5). The segments of a gene include, first, its entire transcribed region—the **transcription unit**. The transcription unit includes the coding sequence in exons, any introns, and the 5' and 3' flanking sequences that extend beyond the ends of the coding sequences and are actually part of the first and last exons. Sequences that lie outside the transcribed region—those that occur close to the 5' and 3' ends of the transcription unit as well as those that occur as much as thousands of base pairs away—but that act to regulate the transcription of the gene are also included in our definition of a gene.

As a general convention, genes are represented diagrammatically from left to right in the direction of transcription. The DNA strand with the same sequence as the mRNA is shown at the top, with its 5' end at the left and its 3' end at the right; for convenience, this is often the only strand depicted. The DNA strand that is complementary to the mRNA is called the template strand. The terms "5' flank" and "3' flank" refer to the nucleotide sequences that, respectively, precede and follow the coding region.

This chapter has described the complexity of individual eukaryotic genes. In the next chapter, we will consider how genes and other DNA sequences are organized within chromosomes. This shift in emphasis is something like considering a complicated modern skyscraper and then going on to think about how the whole city is organized.

CHAPTER 7

The Anatomy of Genomes

A S WE SAW in Chapter 6, gene structure is very different in prokaryotic and eukaryotic organisms. The organization of genetic information is also distinctive in the two groups of organisms. Some differences were already obvious from early work—work done before genetics became a molecular science. Thus, chromosomes are sequestered in a nucleus in eukaryotic cells, but not in prokaryotes. Eukaryotes have multiple chromosomes, each of which has a single, linear, double-helical DNA, whereas prokaryotes have only one chromosome, usually a circular DNA double helix. And, of course, the total number of base pairs is much larger in eukaryotic genomes than it is in prokaryotes (Table 3.1, p. 76). There was good reason to believe that the spacing between genes and the relative locations of related genes also differed between prokaryotic and eukaryotic genomes. Experiments in bacterial genetics (Chapter 4) made it clear that prokaryotic genes are closely spaced around the circular genome, leaving little DNA sequence in between. Furthermore, genes encoding enzymes that function sequentially, or whose activities are otherwise related, are frequently closely linked and form a single transcription unit. For example, the three genes involved in lactose utilization, the gene for β-galactosidase and two others, are part of a single transcription unit. Similarly, five genes encoding enzymes that are required for the synthesis of the amino acid tryptophan from simpler nutrients are adjacent to each other in the *E. coli* genome.

Eukaryotes are very different. Quite apart from the large amount of DNA consumed by introns, eukaryotic genes are separated by long stretches of noncoding DNA sequences. Only very rarely are multiple genes transcribed together in a single mRNA.

Besides the genomic DNA in the nucleus, the cytoplasms of eukaryotic cells contain many **mitochondria**, highly organized structures that also contain DNA (Figure 1.1). Plant cells also contain many **chloroplasts**, another type of cytoplasmic body that contains DNA (Figure 1.1). The mitochondria of plants and animals and the chloroplasts of plants are constructed from special membranes, proteins, and DNA. Within mitochondria, nutrients and oxygen are combined to release usable chemical energy for essentially all cellular functions—from intracellular movement and muscle contraction to the synthesis of proteins and nucleic acids. Chloroplasts contain enzymatic machinery that traps solar light energy and uses it to synthesize cellular molecules that ultimately provide heat, chemical, and mechanical energy for most life on earth. Chloroplasts also generate the gaseous oxygen upon which most living things depend. Both mitochondria and chloroplasts contain characteristic DNA molecules—usually circular double helices. These DNAs encode RNA molecules and some of the proteins essential for mitochondrial and chloroplast structure and function. Other proteins found in mitochondria and chloroplasts are encoded by the cell's nuclear genes.

The genomes of viruses that infect eukaryotes can be either RNA or DNA, either circular or linear (see Chapter 2). In general, their genes are densely spaced, much as genes are in prokaryotes. Coding sequences sometimes even overlap, utilizing either the same, or an alternative, reading frame. There are even instances in which complementary regions on both strands are parts of different genes.

Genome Size

THE FIRST estimates of genome size were obtained in the 1960s and gave paradoxical results. *E. coli*, a typical prokaryote, has a genome consisting of 4.5 million base pairs. This always seemed reasonable, given estimates that there are about 4000 *E. coli* genes with an average size of 1000 base pairs. In comparison, many eukaryotic genomes were found to have a thousand times as much DNA—6 or 8 billion base pairs per diploid cell, equivalent to 3 to 4 billion base pairs for a complete haploid genome (Table 3.1). This was puzzling because it was (and is) estimated that about 100,000 genes should be sufficient to encode a complex eukaryotic organism: This means that, if the average gene is 1000 base pairs long, only about 100 million of the genome's 3 to 4 billion base pairs would be expected to be in genes. Some fungal genomes—including those of yeasts, for example—contain between 10 and 100 million base pairs, but most mammalian and plant genomes are much larger.

Now that large regions of cellular DNAs have been cloned and sequenced, the bulk of the "excess" DNA in eukaryotic genomes can be accounted for. Much of the excess is in introns (Table 6.1, p. 127), and part is in long stretches of additional noncoding DNA between genes. Thus, while genes are stretches of DNA, not all stretches of DNA are functional genes. Additional portions of the apparent excess DNA are explained by multiple copies of some genes.

The existence of multiple copies of particular genes is not an exclusive property of eukaryotes; *E. coli*, for example, has seven genes for rRNA, probably to meet the need for the rapid synthesis of large amounts of ribosomes. Transcription of only a single gene would produce rRNA too slowly to supply the needs of a rapidly growing cell. Eukaryotes also have multiple gene copies to solve similar supply problems; for example, there are several hundred rRNA genes in a typical mammalian genome. But eukaryotes utilize multiple gene copies for other purposes. Different—nonallelic—versions of genes encoding identical or very similar proteins are often expressed in special tissues or at specific times in development by virtue of their association with different regulatory elements; the proteins have identical functions. In other instances, additional copies of a gene may differ enough in their nucleotide sequences, and thus in the proteins they encode, to provide for somewhat different functions.

Yet another portion of "excess" DNA is contributed by multiple, nonfunctional copies of genes, so-called **pseudogenes**. Pseudogenes are usually unable to express useful information because they contain deletions (that is, they are missing portions of their coding sequences), or because mutations have prevented their expression into proteins, or because of the absence of appropriate promoters or regulatory sequences. The number of pseudogenes is different for different genes (from one to one thousand) and, for any particular gene, usually varies from one species to another, being especially high in mammals.

Thus, in eukaryotic genomes, a gene sequence may be represented only once, or it may be part of a family of repeated sequences (Figure 7.1). The members of a family of repeated sequences may include several functional genes that are closely related but are expressed at different times in development or in different tissues or cells, or they may include pseudogenes, or both. They may be clustered together in one region of the genome, or they may be dispersed, even to many different chromosomes.

Still other contributors to the large size of certain eukaryotic genomes are families of sequences that are repeated many times but whose significance is still poorly understood. Molecular cloning and sequencing confirm the

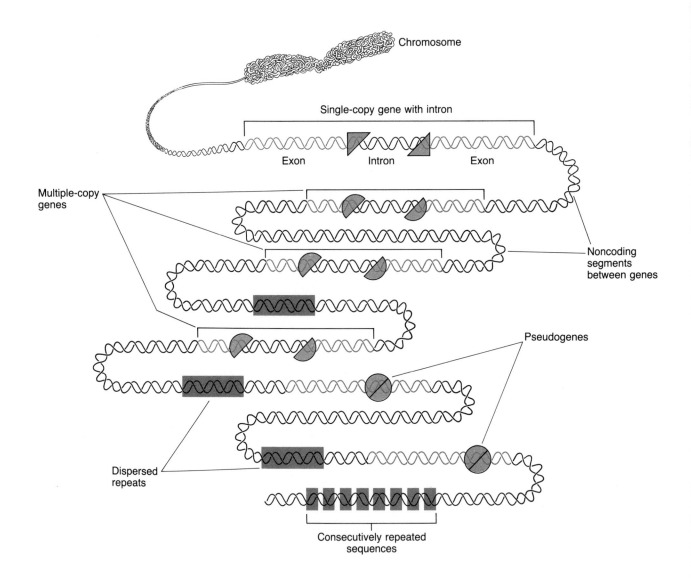

Figure 7.1 Occurrence of different kinds of unique and repeated DNA segments on chromosomal DNA.

existence of such highly repeated DNA sequences, which were first revealed by more classical biochemical techniques. Such families of repeated sequences may contain hundreds of thousands—or even millions—of repeats, and they often constitute 10 percent—and in some instances nearly 50 percent—of a genome. The length of the repeated units varies from two to thousands of base pairs. These repeated sequences, like repeated genes, occur either consecutively (for example, 5'-AGGAGGAGGAGG . . . , etc.) or dispersed to many separate chromosomal locations.

DNA Sequence and Chromosome Architecture

MOLECULAR analysis has revealed that certain kinds of DNA sequences are associated with distinctive structural features of chromosomes, such as centromeres, telomeres, and nucleolar organizer regions. For example, the chromosomal regions referred to as nucleolar organizers contain consecutive repeats of genes for rRNA. The centromeres and telomeres of most eukaryotic chromosomes also contain consecutive repeats of nucleotide sequences. The significance of the centromeric repeats is not well understood. The consecutively repeated sequences at centromeres are different in every species examined. In fact, such repeated sequences may not be essential for centromere function: In at least some organisms, such as yeasts, centromeres function very well without them. DNA sequences representing functional yeast centromeres have been cloned. They are only a few hundred base pairs long and contain no repetitions.

Telomeres coincide with the ends of the long, linear, double-helical DNAs in chromosomes. The DNA sequences at telomeres are very similar in all eukaryotes. They contain variable numbers of short, consecutive repeats, such as

$$5'\text{-CCCTAACCCTAACCCTAA} \ldots$$
$$3'\text{-GGGATTGGGATTGGGATT} \ldots, \text{etc.,}$$

in humans, or

$$5'\text{-CCCCAACCCCAACCCCAA} \ldots$$
$$3'\text{-GGGGTTGGGGTTGGGGTT} \ldots, \text{etc.,}$$

in some protozoans. The telomere repeats are added to the ends of newly replicated DNA by a special enzyme. These unusual terminal sequences

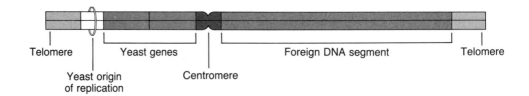

Figure 7.2 Artificial yeast chromosome.

allow cellular processes to distinguish between a true telomere and a broken chromosome. Broken chromosomes are often joined together randomly, but telomeric ends protect chromosomes from such undesirable events.

We have already noted that functional yeast centromeres have been cloned; functional yeast telomeres have also been cloned. Using recombinant DNA methods, linear DNA molecules have been constructed that contain a yeast centromere segment, an origin of replication functional in yeast, a few yeast genes, and the yeast telomeric segments at the ends (Figure 7.2). When introduced into yeast cells, these molecules behave like normal chromosomes in mitosis and meiosis. These constructions are, indeed, artificial chromosomes. It is possible to include very long segments of more than 200,000 base pairs of foreign DNA—for example, human or mouse DNA—in such artificial yeast chromosomes. Thus, yeast cells and artificial chromosomes have become an important host-vector system for cloning large regions of genomes.

Mapping Genes on Chromosomes

NO SIMPLE principles are known to govern the distribution of genes on eukaryotic DNA. Thus, the locations of genes on chromosomes must be determined experimentally. Relatively extensive genetic maps of fruit fly chromosomes were obtained by classical genetic and microscopic techniques in the first part of the twentieth century. These maps, however, accounted for only a small number of the total genes, and, until recently, the genetic maps of most other eukaryotes were even more primitive and uninformative. In humans, several genes were mapped to the X chromosome as early as 1911. The ability to do this depended on the convenient fact that males have only one X chromosome.

Sexually reproducing organisms are generally diploid and contain homologous pairs of chromosomes (Chapter 1). The two chromosomes in a homologous pair look alike, but, more important, they contain very similar DNA double helices and the same set of genes. Any particular gene may be the same on both chromosomes (identical alleles, homozygous) or somewhat different (different alleles, heterozygous). In heterozygous individuals, if one of the alleles is nonfunctional, the organism may still be normal if the second allele encodes a functional protein. This is why mutations are often difficult to detect in eukaryotes but easy to detect in haploid organisms, such as bacteria. But in sexually reproducing eukaryotes, the sex chromosomes—X and Y—do not form a homologous pair. In humans and flies, for example, normal females have two X chromosomes but no Y chromosome; normal males have one X and one Y chromosome. Therefore, a nonfunctional allele of a gene on an X chromosome is readily apparent in males, there being no chance to have a second, normal gene. Red–green color blindness and such inherited diseases as hemophilia and Duchenne's muscular dystrophy, which typically occur in males and are inherited from the mother, were readily associated with the X chromosome (Figure 7.3). The genes responsible for red–green color blindness were the first human genes to be located on a particular chromosome. It took 57 years—from 1911 to 1968—for a human gene to be located on any but the X chromosome.

By 1973, only about 60 human genes had been located on the 22 nonsex chromosomes (autosomes). But by 1981, after the introduction of recombinant DNA methods, the number had grown to 400. By 1991, because of the many DNA segments that had been cloned, the number had risen to about 2100. However, even these greatly expanded maps fail to reveal any systematic placement of genes. Related genes, or genes that are present in more than one copy, may be clustered close together or may be dispersed on different chromosomes.

For example, the human genome encodes five different hemoglobin β-polypeptide chains, each of which is expressed at a particular time during development, from embryo to adult. The genes encoding the five β-poly-peptides (β-globins) are arranged consecutively on chromosome 11. Similarly, the three different genes specifying the α-polypeptides (α-globins) of hemoglobin are clustered, but on chromosome 16 (Figure 7.4). Pairs of α-globins and of β-globins—four chains in all—are required to construct a functional hemoglobin protein molecule that can carry oxygen through the blood from the lungs to all tissues. The different types of α- and β-globins are utilized at different times in the life of the organism. For instance, the α-globin variant called ζ-globin ("zeta-globin") and the β-globin counterpart called ε-globin ("epsilon-globin") are made only in very early em-

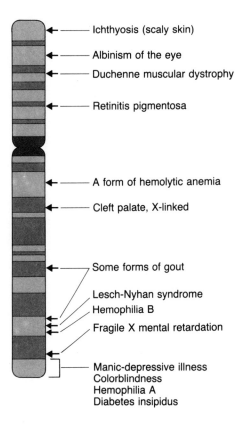

Ichthyosis (scaly skin)

Albinism of the eye

Duchenne muscular dystrophy

Retinitis pigmentosa

A form of hemolytic anemia

Cleft palate, X-linked

Some forms of gout

Lesch-Nyhan syndrome
Hemophilia B

Fragile X mental retardation

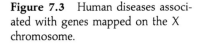

Manic-depressive illness
Colorblindness
Hemophilia A
Diabetes insipidus

Figure 7.3 Human associated with genes mapped on the X chromosome.

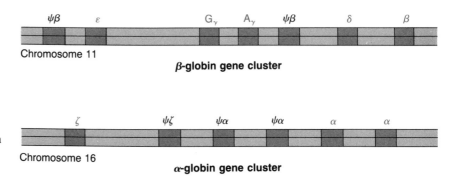

Figure 7.4 Maps of the clusters of human genes that encode hemoglobin polypeptides (the Greek letter psi, ψ, indicates pseudogenes).

bryos. Expression of these genes is switched off later in development, and, during fetal life, the two β-globin variants called γ-globin ("gamma-globin") take over, as do the α-globin variants themselves. Finally, after birth, the δ-globin ("delta-globin") and β-globin genes are predominantly expressed. All the hemoglobin polypeptide genes are related in structure, and one α-globin and one β-globin gene must be switched on if functional hemoglobin is to form. Nevertheless, the two sets of genes are not linked in the human genome (or in other mammalian genomes). How the two sets of globin genes, located on different chromosomes, are regulated in such an exquisite way is an interesting question. In Figure 7.4, notice that both the α-globin cluster and the β-globin cluster contain pseudogenes.

The globin genes were among the first mammalian genes to be cloned by recombinant DNA techniques. More than 90 percent of the protein in red blood cells is hemoglobin, the cells being essentially dedicated to the production of this single protein. Consequently, the bulk of the mRNA in red blood cells encodes the α- and β-globin polypeptides. Because the two kinds of globin mRNAs could be easily isolated, they provided a convenient source of material to construct probes for identifying—in recombinant DNA libraries—clones carrying the corresponding globin genes (the use of such probes is described on p. 109). Some of the clones also contained extensive stretches of DNA that flank the genes in the genome. The flanking DNA segments were then cloned by themselves and were used as probes to isolate, from the library, clones that contained segments of genomic DNA that occur further along on the chromosome. By repeating this procedure sequentially, it was possible to "walk" along the chromosome and reconstruct an entire region encompassing 65,000 base pairs and containing the five human genes of the β-globin family. Subsequently, the entire nucleotide sequence was determined, thus defining the locations of the exons, introns, and regulatory sequences, as well as the locations of members of families of very abundant repeated sequences, which are dis-

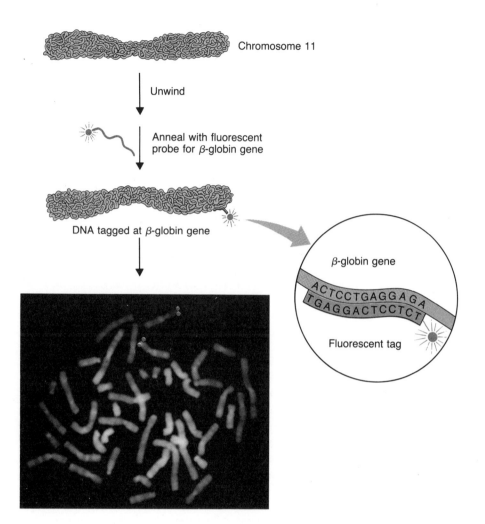

Chromosome 11

Unwind

Anneal with fluorescent
probe for β-globin gene

DNA tagged at β-globin gene

β-globin gene

ACTCCTGAGGAGA
TGAGGACTCCTCT

Fluorescent tag

Figure 7.5 Locating the position of the β-globin gene on human chromosome 11.

cussed in Chapter 8. The complete sequence of the 65,000 base pairs provides the ultimate physical map of this region of the human genome.

On a larger scale of organization, the location of the β-globin gene cluster on chromosome 11 was established by two widely used techniques. In one technique, a DNA probe that contains a fluorescent tag is annealed with mitotic human chromosomes; subsequent microscopic examination of mitotic chromosomes shows the position of the probe as fluorescent spots on a homogeneous stained background (Figure 7.5). In this case, the β-globin probe "lights up" the pair of chromosome 11 and no other. The second technique utilizes laboratory-grown cells that contain a full complement of the chromosomes of one species but only one or a limited number of chromosomes from the second species. For example, if there is a full complement of mouse chromosomes and only human chromosome 11, the human β-globin genes will be found in DNA obtained from the cells. If another human chromosome is present instead, no human β-globin gene is detected. Thus, one can pinpoint the chromosome carrying a particular cloned gene (or other DNA sequence) by annealing the cloned probe to the DNA from a set of mouse cells, each of which has a different human chromosome. The presence or absence of the gene is determined by DNA blotting. For the blotting, DNA is isolated from the cells, and the DNA probe is, for example, a cloned human β-globin gene.

The map of the several β-globin genes in Figure 7.4 is a **physical** or **molecular map**. If all the detail were included, it would show flanking DNA, exons, introns, regulatory sequences, restriction nuclease sites, and, ultimately, the DNA sequence. Such maps are quite different from classical genetic maps, which show the relative positions of genes along a chromosome rather than the molecular detail. Classical genetic maps are deduced by measuring recombination frequencies. Such deductions rely on mutations that give recognizable changes in readily observed organismal structures or functions. In contrast, physical maps can be constructed using cloned DNA segments, even if they are not known to be associated with any organismal trait. Correlations between such physical maps and classical genetic maps can often help to locate the positions of genes on chromosomes. Ultimately, this can lead to cloning of a gene previously known only by its effect on the organism. When the cloned gene is available, the protein it encodes can be synthesized in the laboratory using an expression vector (p. 113), and the significance of the gene and its protein product may then be investigated.

The enormous potential of combined genetic and physical maps was realized in the mid-1970s in studies on viral genomes. This realization led to the current deep understanding of several viral life cycles. Similar information will be critical if we are to understand complex genomes, such as the human genome. There is great interest, therefore, in establishing dense genetic and physical maps for many organisms. Acquiring such data means surmounting various technical and intellectual barriers, but the magnitude of the task is its most challenging aspect.

The first step in establishing a complete physical map is to describe a genome (or its individual chromosomes) as an ordered set of restriction nuclease fragments. A restriction fragment map of the *E. coli* chromosome, which is approximately 4.5 million base pairs long, has already been constructed (Figure 7.6). The entire genome is contained in 21 DNA fragments

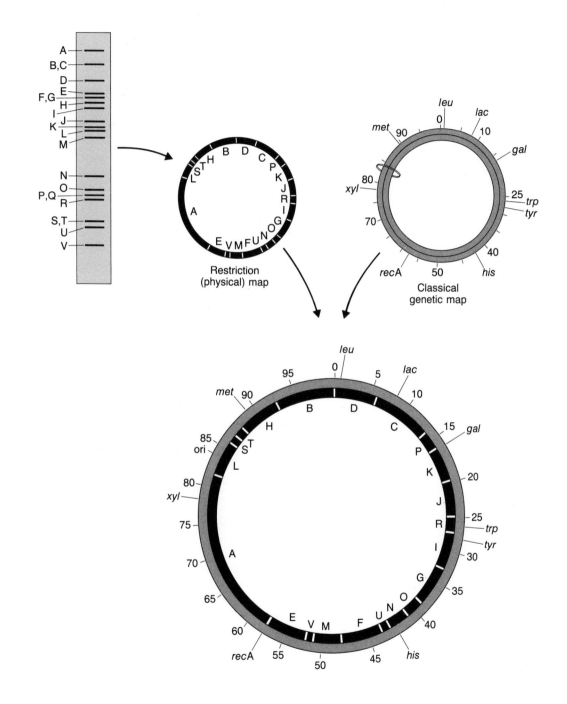

Figure 7.6 Correlating the physical and genetic maps of *E. coli*.

produced by a restriction nuclease that cleaves the DNA relatively rarely. On the map, each of the 21 fragments is indicated by a capital letter. The positions of the fragments on the circular genome were determined by a combination of genetic and biochemical methods. Cloned segments of known genes were used as probes for DNA blotting to determine which fragment carries which gene. Some of these genes are also seen in the genetic map in Figure 4.3 (p. 83). For example, the *gal* gene is on fragment P. Because more than 1000 genes had already been located on the genetic map of *E. coli*, a detailed correlation of genetic and physical maps could be made. Any known or newly discovered *E. coli* gene can now be placed on one or another of the restriction nuclease fragments.

Similar approaches are being used to establish molecular maps for larger, more complex eukaryotic genomes. However, because of their larger sizes, these genomes must first be divided into manageable units. After all, the *E. coli* genome contains only 4.5 million base pairs, whereas many animal and plant genomes contain about a thousand times as many. One approach that is being used is to isolate individual chromosomes and construct their physical maps. Mapping can be facilitated by making a recombinant DNA library with the DNA from an isolated chromosome. The physical maps of the cloned segments can then be made. Clones that contain overlapping segments of the chromosome's DNA can then be used to reconstruct the order of the cloned regions on the chromosome. However, even mapping the DNA in a single mammalian chromosome is a formidable task, compared with mapping the *E. coli* genome. It is helpful if the cloned segments are relatively large. For this reason, the artificial yeast chromosomes mentioned earlier in this chapter (p. 142) are being widely used; segments greater than 500,000 base pairs can be cloned, compared with the maximum of about 40,000 base pairs that can be cloned with phage vectors.

Already, restriction nuclease site maps (that is, physical maps) of extensive regions of two small eukaryotic genomes, those of a nematode worm and baker's yeast, are available and stored in computerized data banks. More than 95 percent of both genomes is covered by partly overlapping DNA segments represented by clones. Experience being acquired with these small genomes is being applied to the mapping of mouse, human, and corn genomes, among others.

The overlapping clones obtained in mapping projects will eventually be used to determine the complete nucleotide sequence of the genomes. Meanwhile, the sequences of segments that have been cloned for specific purposes are accumulating. By May 1992, about 16 million base pairs of sequence from scattered regions of the human genome were recorded in centralized computerized data banks, and extensive data were also available for yeast, fruit flies, and mice.

Merging physical maps with genetic maps can be started, as with *E. coli*, by using genes that have already been cloned and also mapped by genetic methods. For such organisms as baker's yeast, nematode worms, and fruit flies, whose genomes are relatively small (Table 3.1, p. 76), there are comparatively rich genetic maps available. The combined physical–genetic maps make it possible to clone genes that are defined by function and relative chromosomal position, but whose encoded products are unknown. For example, a mutation in fruit flies that has been known for a long time causes an abnormal circadian (24-hour behavioral) rhythm. The same mutation alters the repeat time of a male fly's courtship song. A gene concerned with behavioral periodicity was thus defined. After the gene was located—by classical genetic methods—to a position on the X chromosome, the sequences near this position were cloned. By examining differences in the physical maps of normal and mutant flies, it was possible to locate the gene's precise position and to clone it. Expression vectors can now be used to produce the corresponding protein, and consequently the mechanism behind this interesting and universal behavioral rhythm is being unraveled.

Mapping Large Genomes

EXTENSION of mapping techniques to the very large, complex genomes of plants and vertebrate animals, including humans, is a great challenge. Problems arise, on the one hand, from the enormous sizes of such genomes and, on the other, from the general paucity of classical genetic markers. With humans, the problem is compounded by the unacceptability of subjecting people to experimental breeding or mutagenesis. In the past, human geneticists had to depend on finding rare, large, multigenerational families in which a particular disease occurs with high frequency. Such families are likely to be carrying specific functional mutations, and the pattern of inheritance of the disease can help to determine gene linkages and to establish genetic maps. However, the rarity of such families severely limited what could be learned about human genetics, and the lack of adequate human genetic maps thwarted early diagnosis of genetic diseases. Until recently, reliable diagnostic procedures for fetuses were available for only a few genetic diseases. For example, examination of chromosomes in fetal cells provides a diagnosis for Down syndrome. (This disease is not associated with a mutation, but we classify it as genetic because the affected individuals have three rather than two copies of chromosome 21, as seen in Figure 1.8, p. 22). Another example is Tay-Sachs disease, which is caused by the lack of an essential and readily detectable enzyme. This enzyme prevents the accumulation of toxic products formed during the development of nerve-cell membranes.

Progress with human genetic maps began to be made as larger numbers of cloned segments of human DNA containing genes became available and were mapped to specific chromosomal locations. As described earlier in this chapter, there are two methods that, in conjunction with cloned DNA probes, make the mapping of single-copy DNA sequences on human chromosomes almost a routine procedure. Neither of these methods depends either on mutations or on family trees. Indeed, such mapping can be done even if the cloned DNA probe has no known functions—a so-called **anonymous probe**.

In spite of the large increase in the number of genes and DNA segments that can be located on the human genetic map, the map would have remained sparsely detailed but for new developments. Consider that the 2100 mapped human genes have an average size of 10,000 base pairs (exons plus introns). That accounts for only 20 million of the 3 billion base pairs in the haploid human genome. Any one of the 2000 genes could be separated from its nearest mapped neighbor by many millions of base pairs. At present, about 4000 human genetic diseases are known. Even if the genes associated with each of them could be mapped, we would still be a long way from having the complete genetic map of the human genome. Fortunately, methods developed very recently have begun to give denser and more useful maps for human and other complex genomes, including agriculturally important species, such as corn. These methods too are applicable to humans because they do not require either experimental breeding or large families carrying particular hereditary diseases. They allow the diagnosis of genetic diseases in adults and in fetuses, and they are also useful for studying the distribution of particular alleles in large populations.

RFLPs

THE NEW methods for mapping large genomes depend on the fact that allelic DNA segments on homologous chromosomes frequently have somewhat different nucleotide sequences. They may differ, for example, by a change in a single base pair (for example, from an A · T to a G · C pair) or by deletions or insertions of one or more base pairs, or by rearrangements. Such sequence differences are often detectable by DNA blotting (Figure

5.7, p. 118). For example, the differences may involve a recognition sequence for a restriction nuclease; one allele may have the sequence, the other not. As a result, the number of base pairs between the cleavage sites for that enzyme—in that region of the DNA—will be larger for the chromosome carrying the allele without the site. Consequently, the size of the DNA fragment that anneals with a probe—from that region of the genome— on a genomic blot will be different for the two alleles. Another possibility is that the change deletes the sequence complementary to the probe entirely. In each case, a cloned probe that anneals to the region will give a different pattern, depending on which allele is present. Consequently, the pattern may differ from one individual in a species to another. Even within a single individual, two different patterns can arise from a pair of homologous chromosomes, if they contain different alleles. Fragments of different sizes that are formed from different alleles by a particular restriction nuclease are called **restriction fragment length polymorphisms**, or **RFLPs** (pronounced "riflips"). Analyses of these polymorphisms (literally, "many shapes") can be carried out on the small amounts of DNA available from adult white blood cells or fetal cells that are safely obtained by amniocentesis or by sampling of chorionic villi. Figure 7.7 shows an analysis of RFLPs in the region of a human *ras* gene (the *ras* gene is implicated in a variety of human cancers) among members of an extended family.

One of the surprising findings of restriction fragment length polymorphism (RFLP) analyses is that such polymorphisms are quite common. Already several thousand have been identified. There are many more differences in nucleotide sequences (and thus more alleles of genes) among individuals of a species than might have been expected. Remember, earlier human genetic analysis was based on variant traits that had obvious consequences—for example, eye color or rare inherited diseases. Only a few variant forms of a particular trait were observed, often only one or two. It seemed reasonable to assume that individuals displaying the same variant form, such as individuals with the same inherited disease, had the same alleles. But as the analysis of RFLPs in Figure 7.7 reveals, the number of different alleles of a gene can be large; in the example of Figure 7.7, there are four different alleles. Similar findings have been made in different patients with Tay-Sachs disease; several different aberrant forms of the gene associated with the disease have been identified.

RFLPs occur in both coding and noncoding regions. These findings indicate that there may be considerable diversity among the genomes of individuals of a species. Such diversity can occur within transcriptional regulatory sequences, within introns, and within genomic regions that have

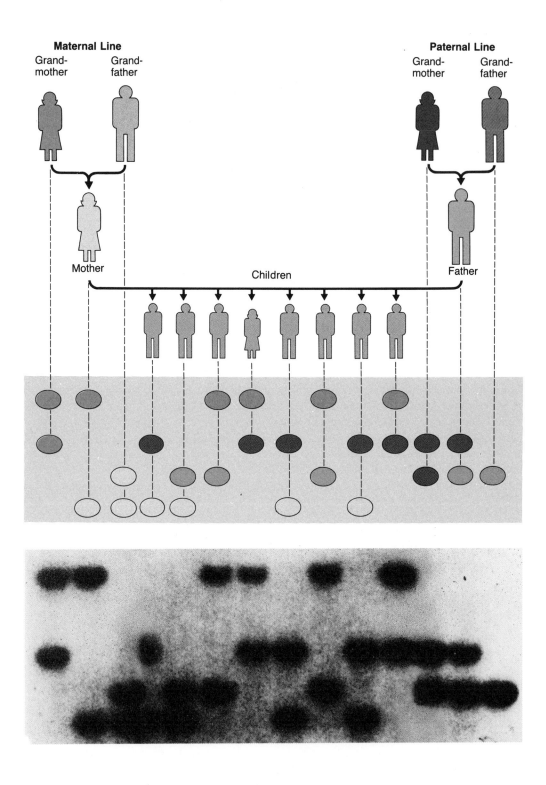

Figure 7.7 (*facing page*)   Inheritance of restriction fragment length polymorphisms (RFLPs) associated with the human *ras* gene. The color-coded diagram provides a key to the pattern of inheritance of four alleles over three generations, as deduced from the photograph of the DNA blot shown below.

no known functions. There may be subtle effects stemming from some of these variations, or many may turn out to be inconsequential.

RFLPs are usually inherited in Mendelian fashion, and thus each RFLP, even if it is detected with an anonymous probe, is, for purposes of genetic analysis, equivalent to a mutation. By relying on RFLPs, which are plentiful, rather than on known mutations, which are rare, geneticists are beginning to fill in the previously sparse human genetic map. The genetic maps of such important plants as corn and tomatoes are also being filled in. Inheritance of a RFLP can be followed by analyzing DNA from members of large, multigenerational families or even population groups. Analysis is not dependent, as it is with mutations, on individuals who display observable genetic abnormalities, although RFLPs may be used for mapping in conjunction with mutations or chromosomal aberrations.

As explained in Chapter 1, recombination analysis began to be used early in the twentieth century to determine linkage of genes and to construct maps of genes on chromosomes. The basic concept can be illustrated by reference to Figure 2.11 (p. 49), which shows meiotic recombination at the DNA level. Three genes, *E*, *F* and *G*, are shown on the two homologous DNA double helices. Each gene has two alleles: *E* and *e*, *F* and *f*, and *G* and *g*. If recombination can occur equally frequently at random positions on the DNA, then the greater the distance between two genes, the more likely it is that they will be separated during recombination. Thus, recombination is much more likely to occur somewhere between *E* and *G* than between *E* and *F*, simply because *E* and *G* are farther apart. If *E*, *F* and *G* are regions containing RFLPs rather than genes, the same is true. Also, *E*, *F* and *G* may be a mixture of RFLPs and known genes. Using different cloned probes, the RFLP method has already defined about 60 positions, spaced, on the average, about 3 million base pairs apart on the moderate-size human chromosome 7.

The RFLP map played an important role in defining the location of the gene that misfunctions in cystic fibrosis, a relatively common human genetic disease. Once the gene was localized to chromosome 7, its precise position was mapped relative to a series of RFLP markers (Figure 7.8). After the gene's position was known, cloning of the gene and characterization of the encoded protein proceeded, leading to an understanding of the defect, new diagnostic tools, and approaches to therapy.

The great interest in RFLPs reflects their importance for the diagnosis of disease. For example, RFLPs provide a straightforward diagnostic tool

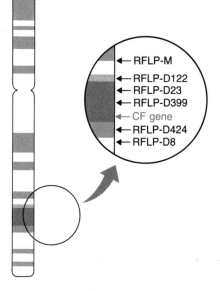

Figure 7.8   Localization of cystic fibrosis gene on chromosome 7 relative to a set of RFLP markers; the markers have arbitrary names.

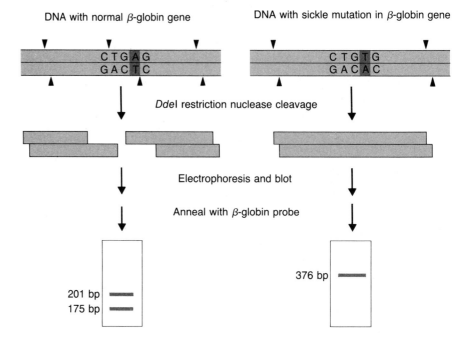

Figure 7.9 Restriction nuclease digests and DNA blotting of human DNA detect the mutation in the β-globin gene responsible for sickle-cell disease.

in cases in which the difference in nucleotide sequence between the normal and mutant allele creates a RFLP. This is true of the mutant β-globin gene that causes sickle-cell disease. Figure 7.9 shows DNA-blot patterns for people with normal and sickle-cell forms of the β-globin gene. The nucleotide sequences of both genes, in the region of the mutation, are also illustrated. The same base-pair change that causes the mutation eliminates a recognition site for the restriction nuclease called DdeI. After normal DNA has been cut with DdeI, electrophoresis and blotting with a radioactive DNA probe complementary to this region of the β-globin gene reveal two radioactive bands. The mutant DNA gives only one larger band, because no cleavage occurs at the absent DdeI site. The distinctive fragment is diagnostic for the sickle-cell gene. It distinguishes normal individuals with two normal alleles from normal individuals who have one faulty and one normal gene (carriers of the sickle-cell gene) and from people who have two faulty genes and will thus suffer from sickle-cell disease.

RFLPs that occur within known mutant genes facilitate diagnosis of defects but, unfortunately, such concurrences are rare. However, a RFLP that maps reasonably close to a mutant gene, even if the RFLP is detectable only with an anonymous DNA probe, can also be used for diagnosis. This is because the RFLP and the mutant gene will tend to be inherited together, only rarely being separated by recombination during meiosis. The closer the RFLP is to the affected gene, the higher the correlation between inheritance of the RFLP and the mutation. Look again at Figure 2.11 (p. 49) and imagine that G is the position of a gene that, when mutated (g), causes a disease. Imagine too that e and f are RFLP alleles of E and F. If RFLP analysis shows that a fetus carries f, it is likely to carry the mutant gene (g) as well. The diagnosis for the disease gene will be less certain if the test was for the e RFLP, because e is further away from the mutant gene than f.

Among the exciting advances from RFLP analysis is the ability to map, diagnose, and identify genes that are associated with poorly understood inherited diseases, including those for which the molecular defect is unknown. To establish diagnostic procedures, large families known to carry the disease in question are screened for a large number of known and mapped RFLPs. The data are searched for statistically significant correlations between the inheritance of a particular RFLP and inheritance of the disease. If the search is successful, the gene associated with the disease can be mapped to a specific region of a particular chromosome. If several known RFLPs flank the gene and are close enough, it may even be possible to clone the gene. For example, the position of the mutant gene responsible for Duchenne's muscular dystrophy was pinpointed by means of RFLP analysis, on the X chromosome. This allowed the normal gene to be cloned. Thereafter, cloned gene sequences were used as a probe against a cDNA library constructed from mRNA from normal muscle cells. Related cDNAs were cloned. The cDNA was inserted in an expression vector, and the encoded protein was synthesized in *E. coli*. (This must be done with a cloned cDNA. The cloned gene can't yield the protein in *E. coli* because bacterial cells have no mechanism for splicing out introns.) A protein with the same properties was then identified in normal skeletal muscle cells but not in skeletal muscle cells from boys with Duchenne's muscular dystrophy. Thus, the disease results from an inability to synthesize the protein, now called dystrophin; this inability is due to a mutated dystrophin gene. This sequence of experiments—leading to the identification of the mutant gene product, starting from the localization of the gene—is being applied to other genetic diseases whose actual causes are unknown. Another successful application of this approach located the cystic fibrosis gene (Figure 7.8). There is now considerable optimism that an understanding of the mutant proteins will lead to therapeutic measures for previously hopeless diseases.

Amplifying DNA Without Cloning

WE HAVE emphasized throughout this book that molecular cloning of genomic DNA accomplishes two major objectives. First, it separates each DNA segment from all others in that genome, and second, it amplifies selected segments of DNA. After a particular gene or region of DNA has been cloned and its sequence is known, certain kinds of mutations within that region can often be detected by changes in the fragment patterns seen following digestion of the DNA with a variety of restriction nucleases, followed by gel electrophoresis, blotting, and annealing with a specific probe. Such is the case for gross alterations in DNA in which base pairs are lost (deletions), gained (insertions), or rearranged. Even a change in a single base pair can be detected by this means, if the change eliminates or creates a cleavage site for one or more restriction nucleases (Figure 7.9). But what about mutations that alter gene function and do not lead to altered fragment patterns following cleavages in that region with a variety of restriction nucleases? How can these be detected without resorting to sequencing of the suspect region?

This problem has been overcome by the development of a new technique called the **polymerase chain reaction (PCR)**. PCR makes it possible to make millions or even billions of copies of any selected sequence in genomic DNA in less than a few hours time. An important feature of this technique is that the DNA segment chosen to be amplified need not be separated from the rest of the genomic DNA prior to initiating the amplification procedure. However, once amplified, the segment can readily be separated from the bulk of the DNA (which is not amplified) by gel electrophoresis. In a sense, the PCR procedure serves in lieu of cloning, because the nucleotide sequence can be determined directly on the amplified segments. Alternatively, sequence changes can be detected and localized by comparing the base-pairing capability of the amplified segment with that of the unmutated sequence using short complementary single-strand DNA probes.

These types of analyses make it possible to screen an individual's DNA for the presence or absence of frequently occurring mutations in known genes. For example, the single base-pair change in the β-globin gene that causes sickle cell disease and the small deletion associated with the predominant form of cystic fibrosis are readily detected in the appropriately amplified DNA segments. The PCR procedure has also greatly simplified the analysis of RFLP patterns for the detection of genetic linkage of selected genomic regions.

The extraordinary selectivity and amplification capacity of PCR is used in conjunction with the analytic procedures mentioned above. This makes it possible to examine a single gene, or a short segment within a gene, even when the total DNA available is that contained in only a single cell. Consequently, PCR is now widely used for detecting virus infections, including the presence of HIV-1 in the blood of patients with (or suspected of having) AIDS; for examining spontaneously arising human cancers to determine possible alterations in the sequences of genes that control cell growth and division; and for typing, prior to organ transplants, the genes that govern tissue compatability. It has also proven to be a powerful tool for the forensic applications described in the next section of this chapter, because the size of sample needed for the analyses is much smaller than that needed for direct analysis.

Figure 7.10 outlines the logic of the PCR method. The principal objective is to copy simultaneously the two complementary DNA strands in a preselected region. To accomplish this, the two strands are first unwound by heating. The single strands are then mixed with a large excess of two kinds of short single strands of DNA, and the mixture is allowed to anneal at an appropriate temperature. Each kind of short single strand forms base pairs with its complementary sequence in one of the two strands of the unwound DNA at the borders of the segment preselected for amplification. These two short single strands serve as initiators for copying the target sequence, each one being extended by the action of DNA polymerase. After this first cycle, each strand of the original DNA segment has been copied once. Now the process is repeated; once again the mixture is heated, causing all the double-stranded DNA to be unwound, and then, by lowering the temperature, the initiator strands still in the mixture anneal to their respective complementary sequences on the original DNA strands as well as on the newly made copies. Once again, DNA polymerase extends the annealed initiators, thereby completing a second cycle of copying. In the third cycle, the same steps are repeated—unwinding, annealing, and DNA synthesis—resulting in further amplification of the region defined by the spacing between the two single-strand initiators. At each cycle, there is a doubling in the amount of double-stranded DNA. Even after only 4 cycles, the segment to be amplified represents more than 50 percent of the total DNA (Figure 7.10). Repeating this process successively for 25 to 40 cycles results in the accumulation of large quantities of the segment chosen for amplification. Millions to hundreds of millions of copies of selected DNA segments a few hundred to a few thousand base pairs in length are generally obtained.

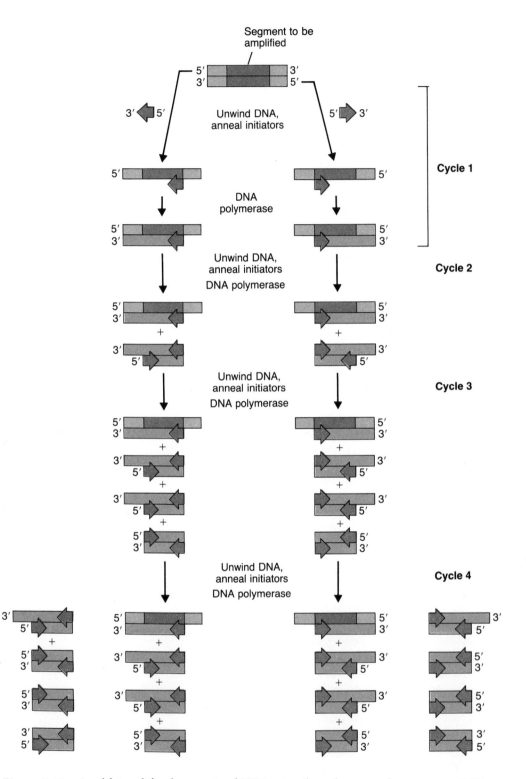

Figure 7.10 Amplifying defined segments of DNA using the polymerase chain reaction (PCR).

The same procedure can be used to amplify—as DNA—a segment of an RNA molecule, if the RNA is first copied into a double-stranded DNA by reverse transcriptase. This latter adaptation of the PCR technique makes it possible to analyze the sequence in mRNAs transcribed from mutant or otherwise interesting genes.

The virtues of speed, simplicity, sensitivity, and selectivity have made PCR an extremely attractive procedure for monitoring the structure of a wide variety of alleles, starting with very small samples of DNA. PCR also simplifies markedly the cloning of rare DNA segments, even genes that are known only from the amino acid sequences of their encoded polypeptides.

DNA Fingerprints

RFLPs reveal a large number of differences in DNA sequences among individuals of a species. But RFLPs of one special kind are even more informative than most. This is because there are an especially large number of different alleles (as many as 50 to 100) at the chromosomal positions associated with these RFLPs.

Consecutive repetitions of particular short nucleotide sequences are found at as many as 100 dispersed positions in chromosomal DNA of vertebrates. The function of these repeats, if any, is unknown. The sequences repeated at these dispersed positions are different from those in the clusters of consecutive repetitions at centromeres and telomeres. Usually, the number of consecutive repeats at any particular dispersed location is relatively small, compared with the millions of repetitions in the centromeric sequences, for example. Furthermore, the number of consecutive repeats at a particular genomic location varies between individuals, thus giving rise to the many distinctive alleles.

Consider what happens when total genomic DNA is cleaved with a restriction nuclease that makes cuts at sequences that flank a group of consecutive repeats but not within the repeats themselves. The mixture of fragments is separated according to size by electrophoresis and the DNA blot is annealed with a DNA probe complementary to the repeats. If the arrows in the diagram below represent the cleavage sites in the flanks, and the numbers each represent one repeat, then a band with general structure

$$-\!\!\downarrow\text{-}1\ 2\ 3\ 4\ 5\text{-}\!\!\downarrow\text{-}$$

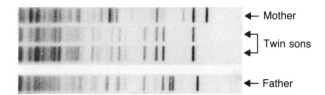

← Mother

Twin sons

← Father

Figure 7.11 DNA fingerprints.

will appear on the blot. An allele that has fewer repeats will give a shorter band:

$$-\downarrow_{-1}\ 2\ 3\text{-}\downarrow_{-}$$

While any particular individual has a maximum of two alleles for any particular genomic location—one on each of the homologous chromosomes—a hundred alleles may exist in the population as a whole. Moreover, the same nucleotide sequence may be repeated at many dispersed chromosomal positions, and sets of differing alleles may occur at all of these positions. All of them anneal to the same probe. The DNA blot will have a band for each position, as seen in the family shown in Figure 7.11. Because of the many positions and the many alleles possible at any particular position, each individual, even in the same family, shows a different **DNA fingerprint.** The mother and father, being unrelated, have completely different DNA fingerprints. The twin sons have identical DNA fingerprints, although, as expected, their patterns include bands inherited from both parents.

When rigorously prepared and analyzed, DNA fingerprints can provide reliable positive identifications of individuals. Almost every individual in a population gives a unique pattern, except for identical twins. A small amount of blood, saliva, or semen, as might be found on the clothing of a victim or suspected criminal, gives enough DNA to carry out the test. Increasingly, DNA fingerprinting is being applied in forensic medicine and to identify lost individuals. And the very same method is proving useful to verify the lineages of costly show dogs and race horses.

The anatomy of genomes appears at first glance to be an arcane subject, likely to be interesting only to biologists. But, in fact, as this chapter illustrates, studying genome anatomy has given us useful tools for medical diagnosis and for breeding plants and animals. Very recent developments indicate that even susceptibility to such complex diseases as diabetes, heart attacks, and Alzheimer's disease will be detectable with these tools. The subject of the next chapter, how DNA sequences rearrange themselves within genomes, may seem even more esoteric. But, as we shall see, it too provides insight into matters of concern to all.

CHAPTER 8

Some Genes Move Around

THE ARRANGEMENT of genes (and most other DNA sequences) within chromosomes is basically stable. Stability is implicit in the ability to establish chromosome maps. Thus, in all individuals of a species, particular genes normally occur on the same chromosome and in the same order. The X chromosome illustrated in Figure 7.3, for example, is typical of all the X chromosomes that are found in the cells of all people. Homologous recombination during meiosis is actually quite common—perhaps as common as one recombination per chromosome at each meiosis. However, this does not disturb the normal alignment of genes and other DNA sequences, because homologous DNA segments are exchanged for one another. Such recombinations bring together new combinations of alleles, but they do not reorder chromosome maps (see Figure 1.10, p. 26). Indeed, many groupings of linked genes are much older than individual species. Thus, genes that are found on the same chromosome in mice, for example, are often similarly linked in humans.

Superimposed on the basically stable organization of genomes is a capacity for the rearrangement of genetic information. Rearrangements that occur at random are generally sporadic and infrequent, many times (but not always) producing deleterious mutations. Thus, chromosomal aberrations, such as breakage, or translocation of portions of one chromosome to a nonhomologous chromosome, have been known for a long time.

In addition to sporadic rearrangements, there are others that are the result of normal processes that occur in specific kinds of somatic cells. These changes alter gene expression in an orderly and programmed manner. DNA reorganization, once thought to occur only very rarely, is now known to be an essential process. The full implications of DNA rearrangement for gene function and for maintaining genome integrity are staggering, if still largely enigmatic. And it is important to recognize that only when rearrangements occur in germ cells can they alter a genome in a heritable way: Changes in other cells are not passed on to offspring.

162

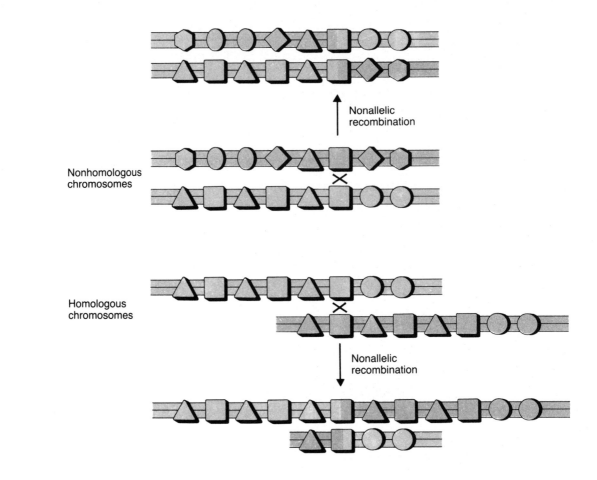

Figure 8.1 Recombination between members of a family of repeated DNA segments that are not allelic creates new chromosomal arrangements. The repeats may be on homologous (*bottom*) or nonhomologous (*top*) chromosomes.

At the molecular level, all rearrangements are changes of DNA structure. All organisms, prokaryotes and eukaryotes alike, encode in their genetic programs enzymatic machinery that can facilitate DNA rearrangements. For example, there are proteins that promote matching of complementary base sequences, enzymes (nucleases) that break DNA strands, and other enzymes (DNA ligases) that make the chemical bonds between nucleotide units.

Various different kinds of rearrangements occur. In one kind, recombination occurs between identical or closely related repeated DNA sequences that are not alleles (Figure 8.1). The sequences can be repeated on nonhomologous chromosomes or at different locations on homologous chromosomes. Unlike the meiotic recombination described at the start of this chapter, such events change the sequences that flank the recombination sites. In another kind of recombination, called **gene conversion** or **sequence**

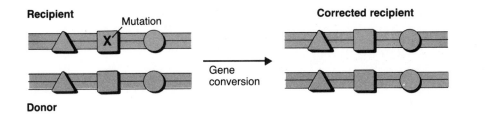

Recipient

Mutation

Corrected recipient

Gene
conversion

Donor

Figure 8.2 Recombination leading to sequence correction: "gene conversion."

correction, a sequence at one genomic location interacts with a homologous sequence—either an allele on a homologous chromosome or a nonallelic repetition (Figure 8.2). One position acts as a sequence donor, the other as recipient. The net result is that a copy of the donor sequence replaces the recipient sequence. A mutation in the recipient segment can be corrected by the donor in this way. The flanking sequences are unchanged. In a third kind of rearrangement, DNA segments move from one position to another; "transpositions" of this sort are described later in this chapter. The formation of additional copies of a DNA sequence, a process called **amplification**, causes a fourth kind of rearrangement (Figure 8.3); the extra copies may be made in place, resulting in a consecutive series of repetitions, or they may be inserted into new locations. Deletion of a DNA sequence is yet

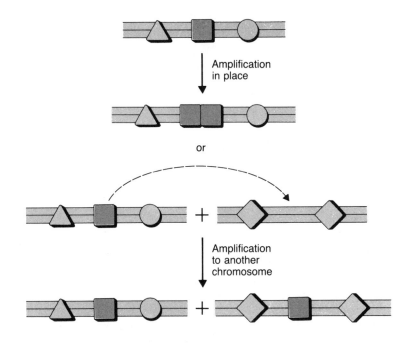

Amplification
in place

or

Amplification
to another
chromosome

Figure 8.3 Modes of DNA amplification

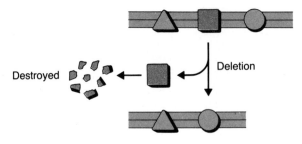

Figure 8.4   Deletion of DNA.

another kind of rearrangement (Figure 8.4), as is the joining of unrelated DNA segments, which can occur when the genomes of infecting viruses are inserted at random positions in a host genome.

The discovery that genomic organization is fluid is one of the major achievements of molecular genetics, and it is especially remarkable because rearrangements of DNA sequences occur very infrequently. To a large extent, this advance stems from experiments that combined the precision of molecular genetics with the highly selective tools of classical genetic analysis, which allow rare events to be observed.

Sporadic Rearrangements

SPORADIC rearrangements are usually random, both with respect to time and with respect to the sequences that are altered. They often produce damaging mutations. However, such rearrangements may also provide, and may have provided in the past, opportunities for novel responses to environmental conditions, thereby facilitating evolutionary change. For sporadic rearrangements to have evolutionary significance, they must occur in germ-line cells or their precursors. Similar events occurring in somatic cells have no effect on the genomes of offspring, and they frequently have little or no effect on the individual organisms in which they occur. However, if the somatic cell is a stem cell—that is, a cell that must give rise to additional important cell types during the organism's development—then random rearrangements may prevent the formation of vital differentiated cells and tissues. In some cases, DNA rearrangements in body cells may lead to tumor formation.

Among the known types of sporadic genomic rearrangements, one in particular—the transposition of DNA segments from one place to another—has captured the imagination of scientists and interested amateurs alike. "Jumping genes" have had the limelight, even in the popular press.

Figure 8.5 Conversion of the smooth trait of peas to wrinkled by insertion of a mobile element into a gene required for starch formation.

The existence of **mobile DNA** elements was first inferred by Barbara McClintock in the 1940s from studies of the genetics of corn (maize). Twenty years later, bacterial geneticists obtained experimental results that led to the same conclusion: Certain DNA segments are able to move about. Ultimately, it was established that many different types of mobile elements occur both in eukaryotic and in prokaryotic genomes.

Molecular cloning has made it possible to determine the nucleotide sequences of many of these mobile elements and to study the mechanisms by which they move to new positions. Mobile elements are genetic troublemakers and cause mutations of various kinds in prokaryotes and eukaryotes alike. When they enter the coding sequence of a gene, the gene's function is lost. When the transposed element lands in the regions flanking a gene, the regulation of that gene's expression can be influenced in complex ways. In fact, many of the mutations that proved so useful early in this century for the study of the genetics of fruit flies are now known to have been caused by transpositions of mobile elements into genes. Even the allele responsible for Mendel's wrinkled peas is likely to be the result of the transposition of a mobile element (Figure 8.5).

Several different types of mobile DNA elements are distinguishable by virtue of characteristic structures or mechanisms of transposition. In general, they are all between 1000 and 10,000 base pairs long. The first

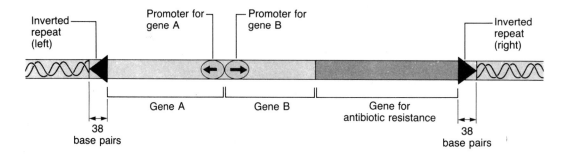

Figure 8.6   A bacterial mobile element.

mobile DNA segments to be studied at the molecular level were those found in bacteria and in their phages and plasmids. Although several different types of bacterial **transposable** (or **mobile**) **elements** are now known, they all share two properties. First, the DNA within the element includes one or more genes that encode proteins required for transposition; in the example in Figure 8.6, these genes are called gene A and gene B. Second, specific, short DNA sequences that are repeated at the two ends of the element are also required for transposition; these are labeled inverted repeat left and inverted repeat right in Figure 8.6. The bacterial mobile elements themselves encode no functions essential for the organisms that harbor them, although they often contain useful genes, such as those encoding antibiotic resistance. On the other hand, these elements can have profound mutagenic effects when they insert in new genomic locations. Transposition appears to occur in one of two ways: Some elements duplicate their DNA during transposition; the original copy remains in place, and a new copy is inserted elsewhere. Other elements appear to be snipped neatly out of one DNA location and reinserted in another, with no duplication of the element DNA itself.

Some eukaryotic mobile elements are like their prokaryotic kin, in that their transposition requires proteins encoded by genes within the element as well as particular nucleotide sequences at the ends of the element. Among these eukaryotic transposable elements are the ones in corn, which were the first mobile units discovered. Similar elements occur in fruit flies, in nematode worms, and in some flowering plants, such as snapdragons (*Antirrhinum* species), where they are responsible for the dramatic and beautiful variegation of petal color (Plate III).

Retrotransposons are another type of eukaryotic mobile element. Transposition of these elements involves the formation of an RNA copy and reverse transcription (thus explaining the cumbersome name). They contain a central DNA segment whose genes encode, among other proteins,

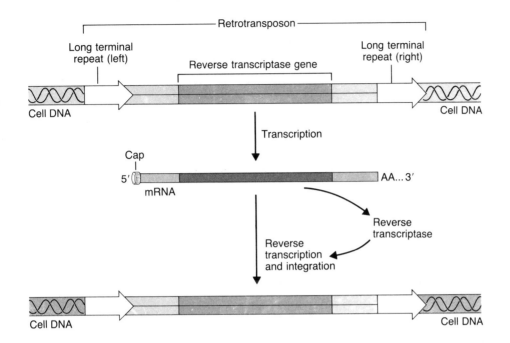

Figure 8.7 A typical retrotransposon.

the reverse transcriptase required for transposition (Figure 8.7). In some retrotransposons, the central segment is surrounded by direct repeats of specific nucleotide sequences several hundred base pairs long—**long terminal repeats**. Such retrotransposons resemble the DNA form of retroviruses, the form that is inserted into host-cell genomes. The mechanism by which retrotransposons transpose is very much like the way in which retroviruses reproduce themselves, a process that is illustrated in Figure 10.5 (p. 201). Unlike retroviruses, however, retrotransposons do not exist outside cells, so they are not infectious. Retrotransposons are known in various eukaryotes, including yeast, fruit flies, and several mammals.

Other kinds of retrotransposons lack long terminal repeats and thus are not like retroviruses. Although less well understood, this second type of retrotransposon also occurs in a wide variety of eukaryotes, including humans. There are almost 100,000 such elements in human DNA. They are counted among the families of dispersed repeated sequences mentioned on page 139. In two boys suffering from hemophilia, the disease was caused by insertion of a retrotransposon into a gene encoding a protein required for blood clotting. Presumably, the transpositions into the X chromosome location of the gene occurred during development of the eggs that formed the two boys. The mothers of the two boys had normal genes in their own somatic cells.

Diverse DNA segments that lack the specific structural and coding properties of mobile elements or retrotransposons also move about in genomes. Unlike the mobile elements, they are highly variable in size and structure. Among these elements are some pseudogenes. Certain pseudogenes derived from genes for various RNAs are remarkably abundant in mammals, with a million copies per genome being not unusual. An interesting family of dispersed repeated elements in the human genome called Alu is in this class.

Each type of mobile element generally occurs many times in the genome. Thus the elements constitute families of repeated sequences that occur at many different chromosomal locations. As such, they foster genomic rearrangements besides transposition, because nonallelic repeats can pair up and undergo homologous recombination. One possible consequence of such recombination between two members of a repeated DNA family on the same chromosome is the elimination of the DNA between them. When such rearrangements occur in single somatic cells in multicellular organisms, they have only a small probability of influencing the function of a complex tissue, but important mutations and chromosomal aberrations can result if they occur in germ cells. Such rearrangements, if maintained within a breeding population, can contribute significantly to evolutionary change.

Human genetic disease can also be caused by the rearrangements fostered by repeated DNA sequences. One such disease, familial hypercholesterolemia, stems from mutations in the gene encoding a particular protein found on many cell surfaces. This surface protein binds the circulating cholesterol, allowing the cell to clear cholesterol from the blood. In diseased individuals, who lack an active form of the surface protein, cholesterol accumulates in blood vessels, causing premature atherosclerosis and heart attacks, even in childhood. Mutations in the gene encoding this protein are relatively common. One in every 500 individuals has one mutated and one functional gene. Such heterozygous people often suffer heart attacks, beginning at about 40 years of age. Individuals having mutations in both genes have the severe form of the disease and, generally, do not survive beyond childhood. One reason for the large number of mutations in the gene is the presence of many copies of the Alu type of short repeated sequence, which occur mainly within the gene's introns. Recombination between two Alus on one chromosome deletes portions of the gene, including coding sequences, making it inoperative. This can be seen in the diagrams in Figure 8.8. At the top is the pertinent 3' portion of the normal gene; the exons are numbered. Note the Alu sequences in one intron and in exon 18. The lower diagram shows the map of a mutant form of the

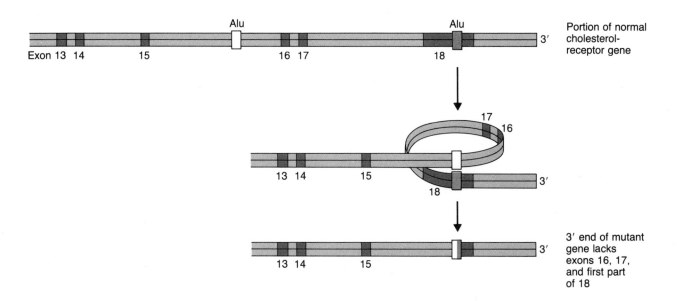

Figure 8.8 A mutation, resulting from recombination between two repeated sequences, in the gene encoding the cell-surface protein that clears cholesterol from blood. The drawing illustrates a possible mechanism whereby such deletions can occur.

gene. The sequences that were between the two Alu segments are lost. Presumably, then, homologous recombination between these two Alus gave rise to the truncated and nonfunctional gene.

It is possible that the kind of human chromosomal aberrations that are visible with a microscope are also generated by such recombinations between repeated sequences that occur at many chromosomal locations.

Programmed Rearrangements

THE EXPRESSION of genes can be profoundly modified by changing their surrounding DNA. A gene may be silent with one set of neighboring sequences but expressed if the flanking DNA is changed. This is one way in which sporadic rearrangements cause mutation. But rearrangements are also utilized in a programmed way to regulate gene expression or to establish specific functions in certain types of cells. Thus, there are genetic programs that cause particular DNA segments to be rearranged at a particular time in the organism's life cycle or in specific cells.

169

Among these programmed rearrangements, none is more interesting or more important for humans than the ones associated with the immune system. These rearrangements are also extremely complex. Only an outline of the basic mechanism is described in the following paragraphs. Keep in mind that, quite literally, our lives depend on these seemingly obscure rearrangements. Later, in Chapter 11, several other examples of programmed rearrangements are mentioned; in all of them, the restructuring of DNA controls gene expression. In the immune system, it does this and more; it actually constructs functional genes.

Antibody Genes

VERTEBRATE immune systems generate a nearly limitless variety of specific proteins. White blood cells—**lymphocytes**—are the synthetic factories that produce these proteins. The role of **immune proteins**—which include **antibodies** and **immune-cell surface receptors**—is to defend the organism against infections by viruses, bacteria, and parasites, and probably also against the growth of tumor cells.

Virtually any foreign molecule can be an **antigen**—that is, can elicit the production of immune proteins. There seems to be no limit to the types of chemical structures that can serve as antigens; even chemicals that do not exist in nature can provoke the production of immune proteins. Moreover, the immune proteins—antibodies and receptors—produced in response to an antigen are highly specific. An antibody against a particular type of bacteria, for example, does not recognize and neutralize other bacteria, even closely related ones. Antibodies against specific chemicals do not "see" closely related molecular structures. A useful (if imperfect) analogy is that of a key (the antigen) fitting into a lock (the immune protein).

Antibody proteins are secreted by lymphocytes and circulate in the blood. The proteins that consitute immune-cell surface receptors reside in the lymphocyte's cell membranes. Remarkably, however, complete genes encoding antibodies or immune receptor proteins do not occur in germ cells or in the early embryo. Instead, the genes are put together later in development, when lymphocytes are formed. DNA segments that are separated by thousands of base pairs in the early embryo genomes are assembled by DNA recombination to form functional genes for antibodies and receptors. Moreover, it is the assembly process itself that generates the enormous number of different antibodies and receptors that can be formed.

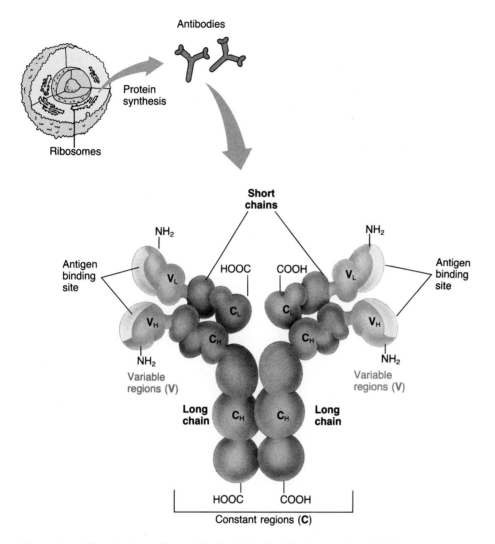

Figure 8.9 The structure of an antibody. Note that the two antigen binding sites are identical.

Figure 8.9 outlines the structure typical of an antibody. Four polypeptides, two short and two long, make up the protein. The two short polypeptides are identical to one another, as are the two long ones. The four are held together by bridges between sulfur atoms in cysteines, amino acids that occur in each polypeptide.

Each of the polypeptides in an antibody protein has two distinct portions, as seen in the antibody molecule in Figure 8.9. The amino acid sequence closest to the carboxyl end is called the **constant** region, because it is nearly the same in all antibodies of a given type (the meaning of the word "type" in this context will be explained later). But the amino acid sequence near the amino terminus is variable, even among antibodies of the same type. It is such **variable regions** that give each distinct antibody—and also each cell-surface receptor protein of the immune system—its unique ability to recognize and bind to a particular antigen molecule. Each antigen binding site (there are two indentical such sites per antibody molecule) is formed from a combination of the variable regions of a short and long polypeptide.

The existence of variable and constant regions in different immune proteins was recognized long before eukaryotic genes could be studied. Many different hypotheses were advanced to explain the enormous diversity of variable regions and the ability of vertebrates to generate antibodies to seemingly unlimited numbers of antigens. At a fundamental level, the problem was how a genome could encode a virtually infinite number of proteins, each with a different variable region but having the same constant region. Also, because the genome cannot know what antigens the organism will need to defend against, how does it know what kinds of variable regions to encode? The answer to both questions is DNA rearrangement.

The general mechanistic features of these rearrangements can be illustrated with one of the simpler examples, the formation of genes that encode one kind of antibody polypeptide—the short chains seen in Figure 8.9. The DNA of a fertilized egg has, on each of one pair of homologous chromosomes, as many as 300 different segments (V) encoding different versions of the variable region that ends up at the amino terminus of the polypeptide (Figure 8.10). Tens of thousands of base pairs away on the same chromosome are several different short DNA segments (J) that encode the rest of the variable region. A few thousand base pairs further on is a segment that encodes the constant region (C) at the carboxyl terminus. As a lymphocyte develops and matures, the DNA rearranges. One of the V segments is joined to one of the J segments, each being randomly selected. The V segment brings with it a promoter. As a result of the rearrangement, a now nearby transcriptional regulatory element in the intron between the J and C segments activates the promoter. Transcription and translation of the rearranged gene leads to production of the short polypeptide, which then associates with a long polypeptide to form an antibody molecule (Figure 8.10).

Because different lymphocytes join different V segments to different J segments at random, each functional lymphocyte can synthesize a dis-

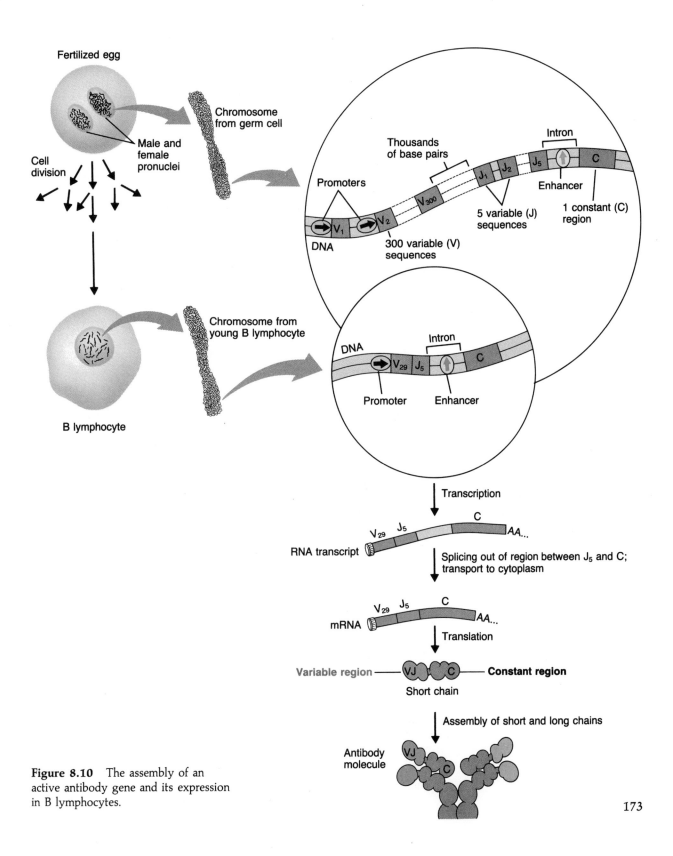

Figure 8.10 The assembly of an active antibody gene and its expression in B lymphocytes.

tinctive immune protein. Considering that any one of about 300 V regions can join to any of the five J regions, about 1500 (300 times 5) different genes can be constructed in this way. Actually, many more different genes are made because V–J joining is imprecise. It yields as many as 10 different base sequences at a V–J junction. Thus, 15,000 (10 times 1500) different genes may result. Because the short polypeptides resulting from such rearranged genes can associate with a large variety of long polypeptides (all of which are encoded by genes constructed by similar kinds of rearrangements) as well as additional means for generating diversity, it is likely that as many as 100,000,000,000 (one hundred billion) different antibodies can be formed in as many different individual lymphocytes.

The rearrangements leading to genes that encode immune proteins occur in developing lymphocytes in the absence of particular antigens. Let us consider one type of lymphocyte, **B cells**. The DNA in a circulating young B lymphocyte has already been rearranged to form a functional gene for an antibody. The antibody protein is being synthesized and is located on the cell surface, embedded in the fatty cell membrane (Figure 8.11). As yet, there has been no encounter with an antigen. When the B lymphocyte happens, by chance, to encounter an antigen that "fits" well into the binding site on the antibody on its surface, it can hold on to the antigen. The binding of the antigen triggers a complex set of activities that result in cell growth and cell division. That activated B cell initiates a clone of lymphocytes. The binding of the antigen also results in the secretion into the blood stream of antibodies with the same binding site and, therefore, the same ability to bind to the specific antigen as the cell-surface antibody.

An important aspect of the immune system is what is referred to as its **memory**. When an organism encounters a particular antigen for the second time, it mounts a more effective and more rapid immune response than it did the first time. Years may intervene between the two encounters. This extraordinary property depends on the formation of special B lymphocytes that "remember" the first encounter. Such **memory cells** are among the progeny a B cell produces after it is induced to proliferate by the tight binding of an antigen (Figure 8.11). Circulating and stored memory cells, unlike the cells that secrete antibody, are long-lived. They retain the entire repertoire of rearrangements and mutations produced earlier. On encountering the same antigen later in life, the memory cells are preferentially activated to proliferate and secrete the same antibody as did their ancestral B cell. This contributes to the improved immunological response. The phenomenon of immunological memory also explains why vaccines protect against specific diseases for long periods of time.

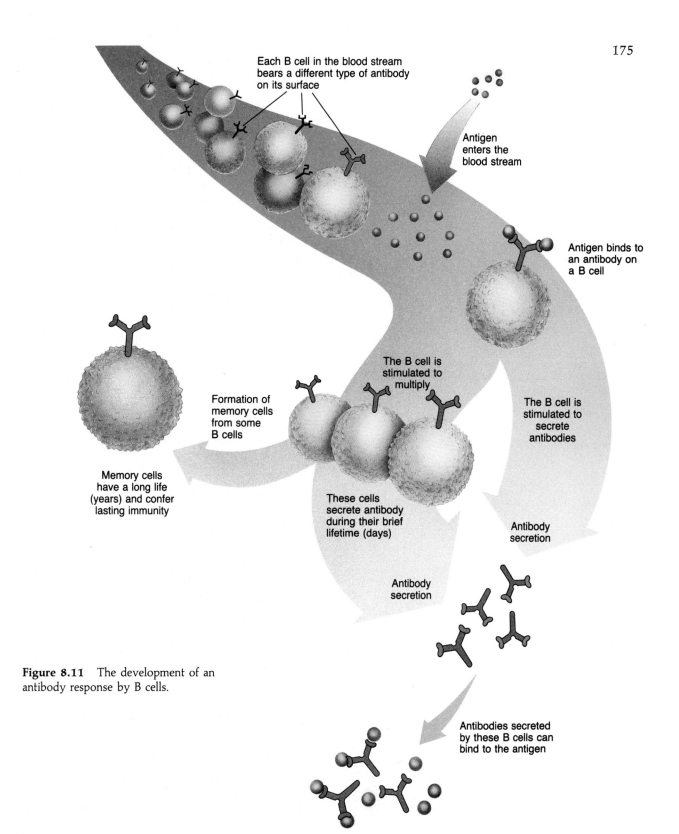

Each B cell in the blood stream bears a different type of antibody on its surface

Antigen enters the blood stream

Antigen binds to an antibody on a B cell

The B cell is stimulated to multiply

The B cell is stimulated to secrete antibodies

Formation of memory cells from some B cells

Memory cells have a long life (years) and confer lasting immunity

These cells secrete antibody during their brief lifetime (days)

Antibody secretion

Antibody secretion

Antibodies secreted by these B cells can bind to the antigen

Figure 8.11 The development of an antibody response by B cells.

176

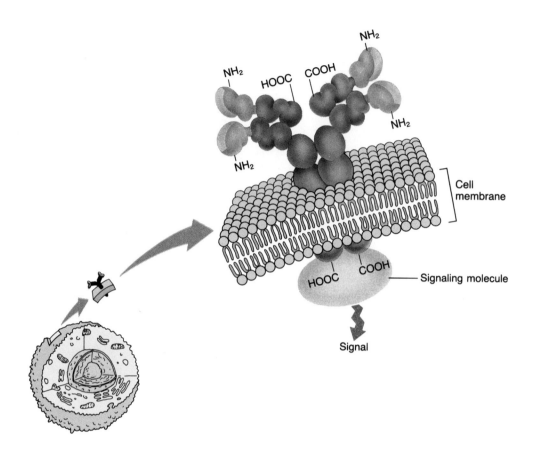

Figure 8.12 An antibody spans the cell membrane of a B lymphocyte.

How does the binding of an antigen on the outside surface of the B lymphocyte trigger activities *inside* the cell? The answer is not known, but it may lie in the fact that the antibody protein actually spans the cell membrane (Figure 8.12). The variable regions near the amino ends of an antibody's polypeptides—which form the binding site for the antigen—are outside the cell; the carboxyl ends of the two long polypeptides are inside the cell. Subtle changes in the molecule's structure caused by the binding of antigen may be "sensed" by molecules inside the cell that can interact with the carboxyl ends. These interactions could then transmit a signal to other parts of the cell, thereby triggering the typical B-cell response. This proposed mechanism is based on known examples of the way small, extracellular molecules convey signals to the inside of a cell when they bind to a cell-surface receptor. The hormone insulin, for example, works in this way.

The processes of growth, proliferation, and secretion of antibody that are triggered in a B lymphocyte by antigen binding are complex responses. They require interactions with other types of white blood cells and with several polypeptide growth factors called lymphokines. Further refinements in the fitness of antibodies for their physiological role as defenders against infections require additional rearrangements of the already formed antibody genes.

Following the activation of B cells, random mutations are induced in the genes encoding the antibody polypeptides. This process alters the amino acid sequence of the variable regions of the antibody. Such changes can improve the "fit" between antigen and antibody, thereby heightening the body's defenses. Moreover they produce even greater diversity in the antibody repertoire. Furthermore, although the long polypeptide of a given surface antibody on a B cell and its corresponding circulating antibody have the same variable regions, they can have different constant regions. These different types of antibodies arise when the antibody gene coding for the long polypeptide undergoes an additional rearrangement, during which the region encoding its carboxyl end (the C segment) is replaced by a sequence encoding a different type of C segment. Thus, antibodies with the same specificity in their antigen binding site can have several different types of C regions. The different C-coding sequences yield different types of antibody proteins for different tasks, such as providing cell surface receptors, producing circulating antibodies, protecting against lung and gut infections, and directing allergic responses.

The freely circulating antibodies secreted by B cells can interact with circulating antigens. The familiar **gamma globulins** (γ globulins) are such circulating antibodies. The antigen–antibody complexes formed by such interactions are then removed by specialized blood cells. Infectious agents such as viruses are also cleared after they encounter antibodies. Young B cells that fail to encounter antigens that fit well into their binding sites die within a few days. New B cells, with unique antibody and receptor genes generated by DNA rearrangement, are formed in bone marrow throughout the life of a healthy mammal.

Among the types of white blood cells required for a B lymphocyte's response to antigen binding are certain T lymphocytes called **helper T cells**. Such T cells also have specific surface receptors. These receptors are also composed of two different polypeptide chains, one short and one long, and these too are held together by bridges between the sulfur atoms in cysteines. Furthermore, each of the two chains in the receptor of a T cell contains distinctive variable regions and a characteristic C region. Specificity for different antigens resides in the receptor's variable regions. Genes encoding the receptors are formed by DNA rearrangements, much like those that assemble antibody genes in B cells, although distinct DNA coding sequences are used. T cells mature in the thymus (which is why they are called T cells), and it is during their passage through this organ that the rearrangements take place. As with the formation of antibody genes, genes for T-cell receptors form in the absence of antigen. The chance binding of a well-fitting antigen to a particular T cell triggers proliferation and the formation of a clone of circulating T cells. All the cells in the clone carry the same receptor, and thus all of them bind tightly to the same antigen.

Other T lymphocytes, **killer T cells,** are the essential components of a separate arm of the immunological defense system. The immune-system receptors on killer T cells interact directly with other cells that carry foreign antigens—for example, virus-infected cells that display viral proteins on their surfaces. This cell-cell interaction triggers a series of events that leads to death of the infected cell.

Our discussion of the immune system began as an illustration of how programmed DNA rearrangements contribute to normal gene expression. This focus, enriched by the detailed molecular explanations acquired in the last decade, ignores other remarkable immunological phenomena. One is the ability of the immune system to distinguish **self**—chemical structures indigenous to an organism—and **nonself**—those structures that are foreign. Rabbits, for example, readily produce antibodies to normal human proteins but not to their own related proteins, even if the rabbit and human proteins differ in only a few amino acids. Rejection of transplanted cells and organs, such as kidneys and hearts, is one consequence of this discrimination between self and nonself, because even individuals in the same species have markedly different versions of certain cell-surface proteins. Autoimmunity, the development of antibodies to an organism's own proteins, occurs rarely, usually in connection with such pathological conditions as juvenile diabetes, rheumatoid arthritis, multiple sclerosis, and certain forms of hypothyroidism. Immunologists are eager to understand why, under normal conditions, antibodies to an organism's own proteins do not develop. If this were understood, it might well lead to effective therapies for autoimmune diseases and ameliorate the problems associated with transplantation of organs. Certain kinds of T cells are known to be important in autoimmunity, but the basic mechanism behind the phenomenon is not known.

Vertebrates, including human beings, are more dependent for good health on their immune systems than on anything else. A severe combined immune-deficiency disease of mice (SCID) results from defective rearrangements of genes for immune proteins both in B cells and in T cells. Similarly, human infants born with certain genetically defective immune systems are doomed to short, disease-ridden lives. Such children are often confined to a sterile plastic tent or bubble, separated as much as possible from bacteria and viruses that children with normal immune systems ward off effectively. More recently, transplants of normal bone-marrow cells, which give rise to normal B and T cells, are being used to treat such immunodeficient children. The devastation caused by the AIDS virus and the challenge to preventing and treating the disease arise because the infection results in a loss of helper T cells, the cells needed to promote a defensive immune response.

The normal working of the immune system is a marvel. It depends on fundamentally random DNA rearrangements. Antibodies with many, many different kinds of binding sites are formed without any "instructions" about what kinds of antigens may appear to threaten the organism. Yet, the diversity of binding sites prepares the organism to defend itself against virtually any invader. Interactions involving B cells and T cells coordinate and improve the defensive reactions.

Luckily, it was not necessary to understand the molecular basis of immune-protein synthesis to develop effective procedures for vaccination. Edward Jenner invented the cowpox vaccination against smallpox in 1796, long before modern immunology began; thanks to this, smallpox was finally eradicated from the planet about ten years ago. A more recent example is Jonas Salk's polio vaccine, which was developed before anything was understood about the genetic basis of the immune response to polio virus proteins. But now with the concepts of molecular genetics and recombinant DNA techniques, new approaches are available for improving the immune response and for designing new kinds of vaccines against previously intractable diseases.

CHAPTER 9

Genes Drive and Record Evolution

THE EVOLUTIONARY histories of organisms are written in their genomes. In the past, evolutionary relationships could be studied only by reference to the consequences of genetic information—the observable traits of organisms. Thus, biological history was first documented by studying the anatomy of fossils and the comparative anatomy, physiology, and embryology of contemporary species. In the middle of this century, when it became possible to determine the amino acid sequences of proteins, comparative protein chemistry was introduced as another indicator of evolutionary relationships. Almost immediately, it became clear that certain proteins that have the same functions in different organisms also have very similar amino acid sequences. This discovery integrated many kinds of organisms into a common evolutionary history in a way that the sparse and accidental fossil record could not achieve. Our present knowledge about DNA structure and genetic processes is completely consistent with the idea of a common evolutionary history for our planet's living things. Even more important, this knowledge is providing clues about how biological evolution occurs. By comparing the structures and functions of genes and the organization of genomes among various contemporary organisms, past events can be deduced, much as a detective can reconstruct a crime from clues left at the scene and an understanding—however sketchy—of human behavior.

Consider first how the comparative analysis of protein structure provides information about evolution. Cytochrome c, for example, is a ubiquitous protein component of mitochondria. The protein is critical to respiration and thus is essential to organisms that use oxygen, including nearly all eukaryotes. The structure of cytochrome c is very similar in all living things that contain the protein. For instance, only 35 out of 104 amino acids are different between the cytochrome c proteins of yeast and humans. In evolutionary terms, this suggests that a cell living more than

1.5 billion years ago, when eukaryotic cells first appeared in the fossil record, had an efficient cytochrome c. That cell passed the cytochrome c gene to its descendents. Some of these cells evolved and diverged, yielding the different evolutionary lineages that led to contemporary fungi, green plants, and animals. Along the way, the amino acid sequences of the cytochrome c proteins in the evolving organisms also changed somewhat, though not enough to interfere with the functions of the proteins. In each evolutionary lineage, the changes were somewhat different, as we see from the contemporary amino acid sequences.

Cytochrome c and other highly conserved proteins are not useful in analyzing the evolutionary relationships between very closely related species because their structures vary little, if at all, among kindred organisms; for example, cytochrome c is identical in humans and chimpanzees. But other proteins evolve more quickly than cytochrome c—for example, the enzyme that catalyzes the simple chemical reaction between carbon dioxide and water to yield carbonic acid. The differences in the structure of the enzyme among primates are sufficient to allow construction of evolutionary trees. This method is based on the observation that the number of amino acid differences between versions of the enzyme in two organisms increases with the time elapsed since the organisms evolved from a common ancestor, as determined from fossil data. In this way, variations in the structure of this enzyme (and of many other proteins) provides a biological "clock."

Finally, however, all methods of evolutionary analysis that depend on organismal traits, such as skeletal features, or cellular traits, such as protein structure, are indirect. Similarities and differences in traits among species depend ultimately on the similarities and differences between their genes and genomes. Thus, the differences in the amino acid sequences of the cytochrome c proteins of yeast and humans depend on differences in the nucleotide sequences of the corresponding genes. The retention (or conservation) of DNA structures dictates the similarities among organisms. Alternatively, the modification of the DNA sequences of germ cells through changes in base pairs or rearrangements underlies the emergence of altered or new traits, and thus of new species, in the course of evolution.

While the study of ancient forms will continue to rely on the morphology of fossils, the comparison of DNA sequences is rapidly supplanting other methods for deducing relationships between contemporary species and their ancestors. Thus, the recombinant DNA techniques and related methods, such as DNA blotting, mapping, and sequencing, have already had a major impact on evolutionary biology, and their influence in this area is likely to be enormous in the future.

Comparing Genes

EVEN BEFORE DNA segments could be cloned and sequenced, the importance to evolutionary studies of comparing DNA sequences was clearly apparent. Methods were developed for assessing the degree of similarity between DNA sequences from different species. These techniques compare the stability of a DNA double helix formed between single strands from two different species with the stability of a double helix in which both strands are from the same species. The basic concept is described in Figure 2.9 (p. 45), except that DNA from two different species is mixed. Strands from two species form imperfect double helices (Figure 9.1). Because of this they come apart at lower temperatures than perfectly complementary strands from a single species. Increasing numbers of mismatches between the two nucleotide sequences make the double helix increasingly unstable, and thus less energy is required to drive the two strands apart. Of course, no double helix forms if the two DNAs have no complementary base sequence.

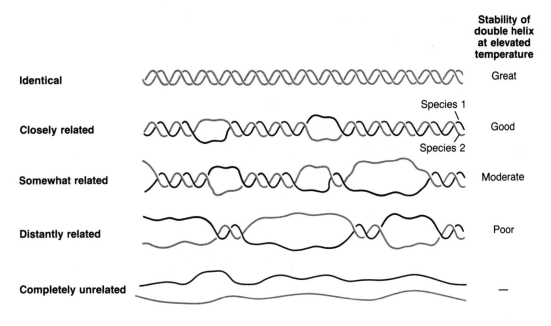

Figure 9.1   Assessing the relatedness of different DNAs.

Useful as they are, studies of this kind give only an overall idea of the relatedness of two DNAs. They cannot approach the richly detailed information of even comparative anatomy and protein chemistry, although they are considerably more rapid. In contrast, comparisons based on direct analysis of the nucleotide sequences in DNA provide fundamental, informative detail. In addition, data about nucleotide sequences in DNA illuminate previously unapproachable aspects of evolution.

With respect to coding regions, DNA sequences are more informative for evolutionary analysis of particular genes than are polypeptide sequences. This is because the genetic code is redundant—that is, most amino acids are coded by more than one codon (see Figure 3.3, p. 64). Consequently, the nucleotide sequence of a gene can be altered without causing any change in the corresponding amino acid sequence of the protein it encodes. For example, although the cytochrome c polypeptides of humans and chimpanzees have identical amino acid sequences, the nucleotide sequences of the genes encoding the two polypeptides are different. And even knowing that an amino acid has been changed during evolution does not necessarily reveal how many base-pair changes occurred. The altered amino acid only allows an estimate of the minimum number of base-pair changes required to substitute one amino acid for another. For example, replacing alanine with valine might occur as the result of changing either one base pair (GCU to GUU) or two base pairs (GCU to GUA). Such information is important if we are to understand the rate at which mutations occur and spread within a population.

A unique advantage of using DNA sequences to study evolution is that noncoding regions, including regulatory sequences, can be compared. This is likely to be of great significance, and it may help us to understand the otherwise puzzling fact that protein evolution and morphological evolution proceed at very different rates. For example, chimpanzees and humans differ markedly in bone structure, hair patterns, behavior, and capacity for language, yet their genome sequences appear to be more than 95 percent similar, and many of their genes encode identical proteins. Perhaps DNA changes that alter the efficiency of transcriptional regulatory sequences cause major differences in the relative rates at which polypeptides are synthesized. This could, in turn, cause profound changes in the rates and patterns of developmental processes. Thus, comparisons of the DNA sequences of regulatory elements or genomic neighborhoods might elucidate the role of gene regulation in the evolution of species.

Genes As History

THUS FAR, we have discussed evolution as it relates to the history of different species. But another kind of biological history, the origin of genes themselves, is also written in DNA structure. DNA sequences reveal a great deal about how new genes, and thus new organismal traits, arise.

Recall that many DNA sequences occur more than once in the genomes of complex organisms. As a result of random aberrations in DNA synthesis or recombination, consecutive repetitions arise. As a result of transposition and recombination, copies of a DNA sequence can be dispersed to new positions, even on different chromosomes. Again, the hemoglobin gene family is illustrative. Multiple genes for β-globinlike polypeptides of hemoglobin occur in a row on human chromosome 11. In addition, multiple genes for α-globinlike polypeptides occur consecutively on chromosome 16. The similarity in coding sequence and intron positions in the family of α- and β-globin genes argues strongly for a common, single ancestral gene. It is highly unlikely that such similarity would have arisen through random processes.

The hemoglobin story is not unique. Two or more related genes (or DNA sequences) within the genome of a single species often derive from a single ancestral sequence. Consider what might have happened when an ancestral DNA segment was duplicated (as a consequence of random events). Assume that one gene copy can supply the required polypeptide (or RNA) in an amount sufficient for normal function. Extra copies are then free from the constraints that limit the accumulation of mutations in a functional gene. Changes in the nucleotide sequence in an extra gene copy could lead to new functional capacities, which might be retained during evolution. Alternatively, an additional copy might acquire new or altered regulatory signals and therefore be expressed at different times during development or only in certain kinds of tissues. Still another possibility is that an extra copy might acquire crippling mutations and remain a pseudogene. The hemoglobin gene family includes examples of all three possibilities. A final possibility is that the redundant gene copy might be lost. This could happen by DNA rearrangement. It could also happen if, by dint of a negative effect of the extra copy or of other mutations, the individual carrying the duplication failed to reproduce.

The previous paragraph describes how new genes can arise from copies of entire preexisting genes. New genes also arise from new combinations

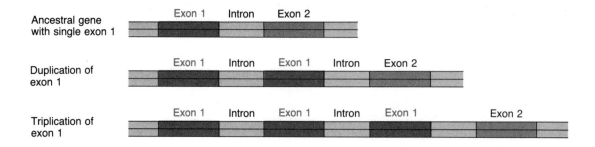

Figure 9.2 Some genes evolved by the multiplication of a single exon.

of copies of coding regions—that is, of exons. In some cases, a single exon can be duplicated to produce a repetition of exons (Figure 9.2). The structure of the new gene is then reflected in multiple repeats of a polypeptide segment in the encoded protein. The gene for the receptor that clears blood cholesterol, for example, contains multiple repeats of exons also contained in other genes. In other cases, genes appear to be mosaics that were constructed by patching together copies of individual exons recruited from different genes—**exon shuffling** (Figure 9.3). In fact, exons with the same ancestor occur in several apparently unrelated genes. Exons that are thus shared by several genes are likely to encode polypeptide regions that endow the disparate proteins with related properties. For example, the compound adenosine triphosphate, a relative of the adenine found in DNA and RNA, binds to many different proteins in order to carry out its function as the central transmitter of chemical energy in all cells. In many proteins, the regions that bind adenosine triphosphate are very similar and are encoded by exons that were probably derived from a common ancestor. Similarly, the repeated exon in the gene encoding the cell-surface receptor that clears blood cholesterol also occurs in genes encoding other cell-surface receptors for quite different circulating molecules. In one sense, this is like the assembly of computers with different capabilities by using a relatively limited number of different kinds of integrated circuits in a variety of configurations.

There is an interesting hypothesis concerning the assembly of new genes by exon repetition and exon shuffling. It proposes that, during evolution, DNA rearrangements occur within introns, thereby linking together intact exons in new combinations. This view implies that introns are ancient structures and were present in the earliest genes. One corollary of this hypothesis is that introns were lost during the evolution of prokaryotes. A contrasting hypothesis holds that introns represent insertions into preexisting coding regions by a mechanism that is efficient in eukaryotes

Ancestral genes

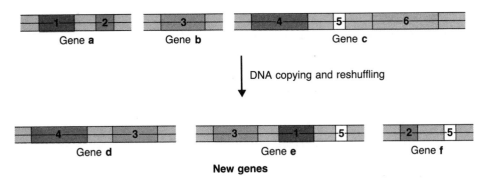

Figure 9.3 Some genes evolved by duplication and shuffling of exons.

but not in prokaryotes. The idea that introns are ancient avoids several problems inherent in the insertion model. First, widespread random insertions of noncoding segments into coding regions are likely to have significant mutational consequences. Second, if introns were introduced late in evolution, it would be necessary to evolve, simultaneously, the complex and precise splicing mechanisms. In contrast, if introns were present in the earliest genomes, then splicing too might be a very ancient process. The discovery that some introns catalyze their own removal by splicing, without the aid of enzymes or other proteins, enhances the plausibility of the very ancient existence of introns. Nevertheless, it is likely that some introns were introduced by insertion; the two hypotheses are not mutually exclusive.

Molecular analysis of DNA sequences reveals that the processes of gene duplication, exon repetition, and exon shuffling were not rare or occasional contributors to genome evolution. The histories of substantial portions of eukaryotic DNA, coding and noncoding sequences alike, reflect such processes. Contemporary genome structures reflect ancient processes that were, and presumably still are, major contributors to evolutionary change. Through such processes, new and occasionally useful capabilities become available to organisms. When such capabilities provide even subtle advantages to an individual in a particular environment, the trait may spread successfully to succeeding generations—the process called **natural selection**. When new traits resulting from new alleles become widespread in a species, they might foster the reproductive isolation of a group of individuals that contain the new alleles and thereby lead to the formation of a new species. Molecular genetics gives new insights into the genetic variation that provides the raw material for evolution. Variation stems not only from small changes in DNA, the classical notion of mutations, but from a host of transactions that remodel genes and genomes.

Origin of Genes

WHAT does the proposed great age of introns imply about the nature of the earliest living cells and their genetic systems? In particular, how does it bear on the question, Which were the first informational molecules—DNA or RNA? Several lines of evidence suggest that RNA came first and provided the basis for the earliest coding systems. For example, rRNA (ribosomal RNA), tRNA (transfer RNA) and mRNA (messenger RNA) are central elements in the machinery all organisms use to translate the genetic code. These molecules must have been present in the earliest genetic systems, predating the evolutionary separation of the prokaryotic and eukaryotic lines. Indeed, the structures of rRNA and tRNA are similar in all organisms. Laboratory experiments show that new, short RNA molecules can be made by copying from an RNA template by purely chemical reactions, in the absence of any enzymes or other proteins. It seems possible then that RNA molecules might have been self-replicating, albeit inefficiently so, and have been transcribed and translated by primitive mechanisms. In addition, RNA molecules can even act like enzymes to modify RNAs. Thus, in *E. coli*, a special nuclease responsible for the cleavage of long RNA molecules that are the precursors of mature tRNAs is made up of a protein and an RNA chain; the RNA chain alone carries out the cleavage. And, as already mentioned, introns in rRNA precursors in certain protozoans and fungi are spliced out without the intervention of proteins. In contrast, DNA is not known to carry out any of these reactions.

The common ancestors of contemporary genetic systems may, therefore, have utilized informational RNA molecules. In these RNAs, short, fortuitously formed coding regions (primitive exons) would have alternated with noncoding regions (ancestral introns). Spontaneous splicing of such introns would result in the joining of the exons to produce continuous coding regions. In time, if short polypeptides that facilitated RNA replication and processing evolved, they would have given the cells that contained them a large reproductive advantage. One of these polypeptides might have had the properties of a primitive reverse transcriptase, allowing it to catalyze the formation of DNA molecules. Later, the DNA could have evolved to become the fundamental repository of cellular genetic information. DNA molecules that had evolved in such a way would have retained the exon–intron pattern of the RNA from which they were copied.

A further extension of this model for the origin of genetic systems imagines that a cell containing DNA became the progenitor of subsequent successful organisms. Separate evolution of different daughter cells could have given rise both to eukaryotic cells, in which retention of introns was favored, and to prokaryotic cells. According to this model, it would not be too surprising to find an occasional intron remaining in a prokaryotic genome. Indeed, introns do occur in polypeptide genes in the genome of an *E. coli* phage, as well as in tRNA genes in the group of organisms known as archaebacteria.

One of the most provocative aspects of the model just described is its implication that prokaryotes and eukaryotes evolved independently from a common ancestral cell. Previously, most speculations regarding early evolution assumed that eukaryotes evolved from prokaryotes. That assumption is based on the idea that the relatively simple contemporary prokaryotic cells are likely to resemble the earliest cells. Larger and more complex eukaryotic cells are then assumed to be the product of evolution from the simpler, prokaryotic types. The discovery of introns, splicing, cellular reverse transcriptases, and the catalytic properties of RNA has forced a reevaluation of such concepts. Thus, our ability to analyze the molecular aspects of biological information systems is shaping new approaches to major questions in evolutionary biology.

CHAPTER **10**

Viruses and Cancer

R ECOMBINANT DNA methodologies and related techniques, including DNA sequence determination, PCR, and restriction-nuclease
analysis, provide rich molecular detail about genes and genomes.
The information confirms the fundamental unity of genetic structure, organization, and expression among different organisms. It also hints at the
enormous variety of biological mechanisms that are used to convert genomes into successful organisms. In fact, the number of genetic strategies
employed by living organisms seems virtually limitless. The principal gen-
eralization is that anything that works is likely to be adopted sometime
during evolution. However, we have only begun to understand the varied
genetic tactics that account for the diversity of organismal form, habitat,
behavior, and function. To extend our understanding, we will need to
analyze complex interacting sets of genes and gene products. In this and
the next chapter, we describe how biologists are analyzing such genetic
systems in viruses (this chapter) and diverse cells and organisms (Chapter
11). The description of viruses leads us directly to the subject of cancer
and the extraordinary new understanding of tumor cells provided by molecular genetics.

A World of Viruses

T HE ULTIMATE goal of understanding the relation between a complete
genome and the organism it produces will be achieved only in the
distant future. However, such an understanding has been approached with
certain viruses. A virus generally consists of a genome enclosed in a protein
coat (Figures 10.1 and 10.2). In some complex viruses, a fatty membrane
studded with proteins surrounds the coat. These shells probably protect
the viral genome as it passes from cell to cell through fluids or air, but the
coat proteins also have an important role in facilitating the entry of the
virus into susceptible cells.

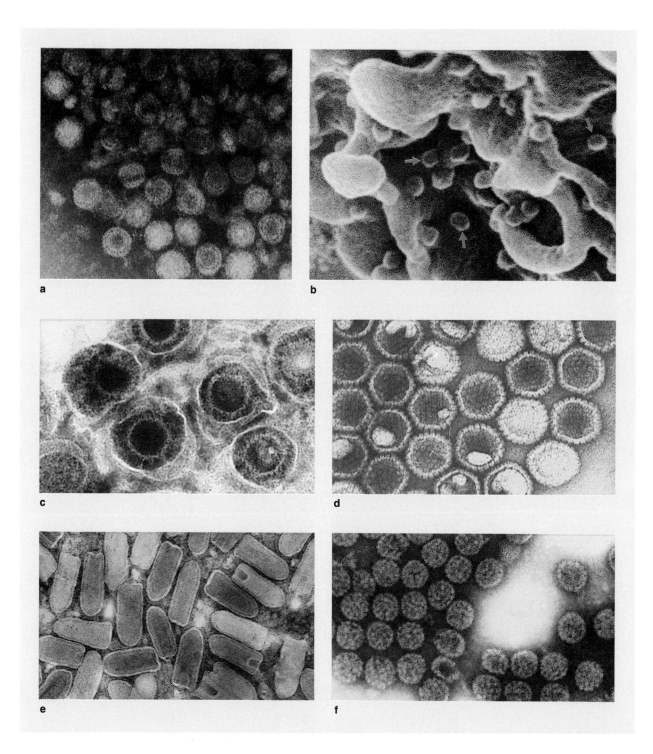

Figure 10.1 Electron micrographs of several eukaryotic viruses magnified about 100,000 times: *a*, hepatitis B virus; *b*, AIDS virus (HIV); *c*, herpes simplex virus, type 1; *d*, herpes simplex virus, type 1, with envelopes removed; *e*, vesicular stomatitis virus; *f*, human papilloma virus; *g*, influenza virus; *h*, a mouse retrovirus.

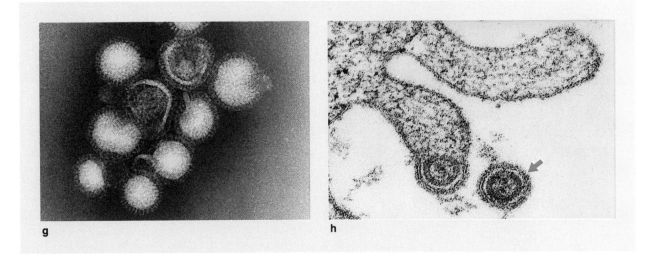

g

h

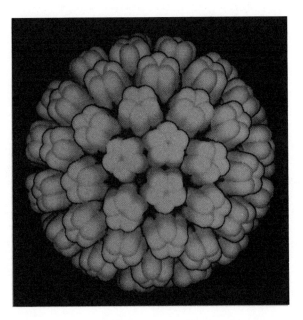

Figure 10.2 Computer-graphics representation of a papovavirus protein coat magnified about 1 million times. (Plate IV is a full-color version of this image.)

Recall that viruses cannot multiply outside living cells: The host cells are needed to provide energy and small molecules, such as amino acids and nucleotides, for making viral proteins and nucleic acids. Moreover, viruses co-opt the host cell's machinery for transcription, translation, and replication in order to express the virus's genetic information. When viruses infect new cells, the coat is usually discarded and the genome begins to direct the synthesis of new virus particles. Virus genomes are relatively small, compared with prokaryotic and eukaryotic cells. They encode a limited number of proteins, varying from as few as five to as many as hundreds, depending on the virus. Among the proteins formed, some will be used to construct the coat, and, in some viruses, some of the virus-encoded proteins are RNA or DNA polymerases. These proteins are needed because cells do not contain the necessary RNA polymerases for replicating viral RNA genomes, nor do they have the proper DNA polymerases or accessory proteins required for replicating certain viral DNAs. We have detailed molecular descriptions of several viruses and their life cycles because of two technical developments: recombinant DNA techniques and sophisticated methods for growing cells in laboratory dishes. The cells allow viruses to be studied without the complications inherent in using whole animals or plants.

Viral genomes may be either DNA or RNA, and either linear or circular molecules. Some viruses kill their host cells in the course of replicating themselves. Others, such as the retroviruses, insert their genomes—or copies of their genomes—into the host-cell DNA, thereby becoming permanent genetic residents of the infected cell and all its offspring. In many cases, the viral genomes in the cell's chromosomal DNA retain the ability to produce new viral particles. This diversity of form and function illustrates the general notion, expressed before, that, during evolution, many different genetic mechanisms are tested, and anything that works is likely to be adopted. In this feature, viruses are no different from complex organisms; they are only smaller and less independent.

DNA Viruses

REPLICATION of viral genomes follows many different routes and depends on whether the genome is RNA or DNA, circular or linear. Although much remains to be learned about virus replication in infected cells, there are already drugs that interrupt the replication of some disease-causing viruses. Among these, AZT (azidothymidine) and ddI (dideoxyinosine) are being widely used to lessen the effects of HIV, the virus that causes AIDS. Gancyclovir is used to combat herpes virus infection. Other drugs aimed at blocking steps in the life cycles of viruses are being sought to provide effective cures for viral diseases of plants and animals. However, immunological prevention by vaccines remains the most effective defense against many virus infections in vertebrates.

The genomes of many DNA viruses are replicated by DNA polymerases according to the same general mechanism used for cellular DNA. These viruses include the papillomaviruses (some of which cause warts), adenoviruses (which are associated with a variety of mammalian respiratory infections), and herpesviruses (one variety of which is responsible for serious genital lesions).

Other DNA viruses, including hepatitis B virus and cauliflower mosaic virus, replicate their genomes in two steps: First, their DNA is transcribed into an RNA by a cellular RNA polymerase, and then the RNA is reverse-transcribed to yield a new DNA genome. In both examples, the reverse transcriptase is encoded in the viral genome. Retroviruses—including HIV, the causative agent of AIDS—have RNA genomes and also encode reverse transcriptase. Other RNA viruses, such as poliomyelitis virus and foot-and-mouth disease virus (which causes a serious and highly contagious disease among cloven-hoofed mammals), replicate by copying RNA directly into RNA, using virus-encoded RNA polymerases.

Recombinant DNA techniques have been especially important in the study of hepatitis B virus, a serious cause of human disease worldwide. In this case, alternative approaches were extremely limited because, until very recently, no laboratory-cultured cells were available in which hepatitis B virus could be grown. Studies had to be carried out on viruses obtained from infected humans. Moreover, only small quantities of the virus's circular DNA genome could be isolated from blood. The amount is, however, sufficient to permit cloning the DNA, thereby providing ample viral DNA for characterization. Viral DNA in an expression vector produces virus-encoded proteins usable for the development of vaccines. The coat protein of hepatitis B virus, synthesized in yeast cells, was the first vaccine produced by recombinant DNA techniques to be licensed for human use in the United States. Similar approaches have also yielded clones encoding a surface protein from the hepatitis C virus, a particularly serious human pathogen. This advance may lead to an effective vaccine against this type of infection. And the race is on to use engineered versions of HIV genes to make proteins that could be used to immunize people against infection by that virus.

Simian Virus 40

SIMIAN virus 40 (SV40), a member of the papovavirus group, was first isolated from monkeys, but it and other papovaviruses can infect humans. The events following SV40 infection are part of an orderly, timed series of processes leading to either of two possible outcomes. Thus, SV40 multiplies in and kills primate cells; but, when it infects rodent cells, SV40 inserts its genome into the infected cell's DNA, often transforming that cell into a cancer cell. In the course of studying these two virus "lifestyles," many of the fundamental regulatory signals governing the transcription of eukaryotic genes were discovered. For example, studies with SV40 promoters revealed for the first time the structural complexities of eukaryotic promoters. Similarly, comparisons of the mRNA and genomic sequences of SV40 led to the discovery of introns and splicing. Today, this virus remains an attractive model system for the analysis of additional questions about the molecular genetics of eukaryotes. It is also the source of critical DNA segments used in the construction of eukaryotic vectors. In addition, the regulatory mechanisms governing virus multiplication have provided clues for understanding differentiation and development in complex organisms.

A closer look at SV40 is worthwhile, because it embodies, in a simple format, many typical viral features. The virus itself appears to be spherical, although it is actually a polyhedron with 20 faces (icosahedron). Its genome consists of a single, circular, double-stranded molecule of DNA represented in grey at the center of Figure 10.3. The DNA is folded into chromatin with the aid of the same proteins used by the host for packing its own genome into chromatin. The entire nucleotide sequence of the DNA (5243 base pairs) is known, and the coding regions for five polypeptides have been identified as illustrated by the colored bars in Figure 10.3: the three different proteins that make up the virus coat (VP1, VP2, and VP3) and two others, called small-T and large-T. After entry into a cell, the virus travels to the nucleus and loses its protein coat (Figure 10.4). The genome is expressed into mRNAs and proteins in a precisely ordered sequence, and, in primate cells under laboratory conditions, new viral particles begin to appear about 30 hours after infection. After four days, all the cells succumb to the infection and burst. Tens of thousands of new viral particles are released into the surrounding fluid from each infected cell.

The earliest detectable events after infection are transcription and translation of the genes encoding the small-T and large-T proteins. Early tran-

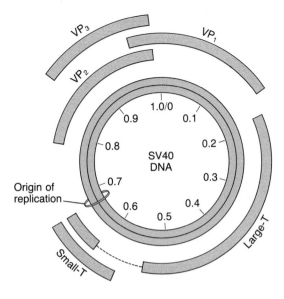

Figure 10.3 The genome of simian virus 40. The five colored regions encircling the genome show the segments encoding the five viral proteins. Several genes overlap.

scription starts at a virus promoter and produces mRNAs encoding the small-T and large-T proteins. In primate cells, viral DNA replication begins when the newly made large-T protein binds to the nucleotide sequence at which replication begins (the replication origin). The large-T protein is needed to begin the replication, the rest of the process being carried out by a battery of enzymes normally employed in replicating the cell's DNA. DNA replication proceeds from the origin in both directions around the circle, and terminates when the copying process reaches halfway around the genome. At that point, the newly made strands and the original viral template strands are joined to form two circular double-helical DNAs. The role of the small-T protein in virus multiplication is still unsettled. As the number of progeny DNA molecules increases, the synthesis of small-T and large-T proteins diminishes, and the genes encoding the coat proteins begin to be transcribed and translated. The resulting accumulation of viral proteins and progeny DNA molecules culminates in the assembly and release of new viral particles and death of the host cell.

We mentioned earlier that infection of rodent cells by SV40 leads to an outcome different from the one just described for primate cells (see Figure 10.4). The initial entry of the virus into the nucleus of a rodent cell leads to the transcription and translation of the large-T and small-T genes, as occurs in primate cells. In rodent cells, however, the large-T protein is unable to trigger replication of SV40 DNA, and, consequently, virus mul-

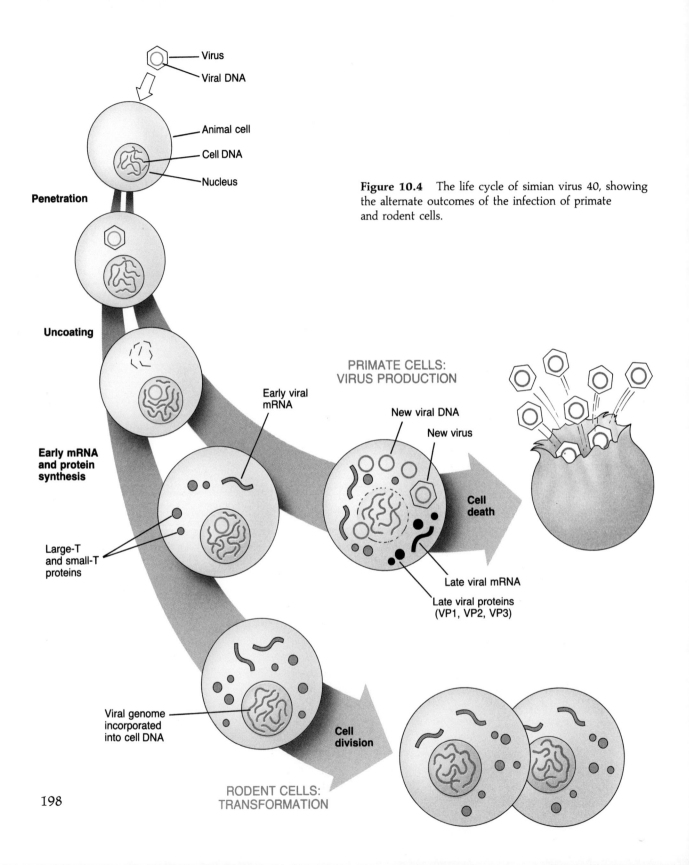

Figure 10.4 The life cycle of simian virus 40, showing the alternate outcomes of the infection of primate and rodent cells.

198

tiplication is aborted. The ultimate outcome of the infection is decided, however, by whether the SV40 DNA, which has reached the nucleus of the infected cell, is lost or inserted into the DNA of the host cell during subsequent growth. If the SV40 DNA is lost, there is no apparent effect of the infection on the surviving cell. But if the viral DNA is integrated into the cell's DNA, the large-T and small-T genes are expressed. In a way that is still not fully understood, the accumulation of the large-T protein in the nucleus alters the cell's growth-control mechanisms, causing it to behave like a cancer cell. In fact, the large-T gene was one of the first recognized cancer genes (**oncogenes**), and the mechanism by which it causes cancer is still being actively sought after.

In contrast to eukaryotic cellular genomes, there is little or no space between genes on the SV40 genome; coding and regulatory sequences overlap (Figure 10.3). For example, the beginning of the large-T and small-T proteins have the same amino acid sequence while their carboxyl ends are different. This is because the RNA transcribed from this region of the SV40 genome is spliced in two different ways to yield two distinct mRNAs (Figure 10.3). Similarly, the coding sequences for the coat proteins VP1, VP2, and VP3 overlap but are in different translation frames.

Adenovirus genomes are seven times the size of SV40 DNA. The enlarged coding capacity permits a more complex life cycle and a more complex structure, although adenoviruses are still comparatively simple: a linear, double-stranded DNA genome surrounded by a protein coat. Here too, important genetic information for producing and regulating virus proteins is closely packed in the viral DNA. Thus, an adenovirus encodes its genes on both DNA strands, and, in some cases, the regulatory and coding sequences overlap. In the large family of herpesviruses, the virus coat is surrounded by a membrane that is studded with proteins encoded by the viral genome. The herpesvirus particles contain about 30 different virus-encoded proteins, compared with three in SV40 and about ten in adenoviruses. The infectious cycles of herpesviruses are quite complex, but their regulation and the proteins that are formed during their course are encoded in a linear, double-stranded DNA genome roughly 150,000–250,000 base pairs in length. The family of herpesviruses includes several that are associated with serious medical and veterinary problems. For example, human herpes simplex viruses 1 and 2 cause recurring facial and genital lesions, respectively; cytomegalovirus often causes a fatal disease in developing fetuses; Epstein-Barr virus is associated with Burkitt's lymphoma and infectious mononucleosis; and Marek's disease virus causes tumors in birds.

RNA Viruses

THE APPLICATION of recombinant DNA technology has also been of special significance to the study of RNA viruses because the sequences of large RNA molecules are virtually impossible to determine directly. Moreover, the DNA form of cloned viral genes, when present on appropriate recombinant vectors within bacterial and animal cells, can be used to produce viral proteins for use in biochemical studies and as antigens for immunizations. For example, the virus that causes foot-and-mouth disease is a formidable experimental challenge because its RNA genome contains 8000 nucleotides. Its importation into the United States is strictly prohibited. However, cDNAs encoding viral coat proteins have been synthesized and cloned, and they can be safely manipulated in the laboratory. Expression of these cloned foot-and-mouth virus sequences in bacteria provides polypeptides that have been used to develop vaccines. Also, because the cloning allowed the amino acid sequence of the coat protein to be deduced, it has been possible to investigate the use of chemically synthesized polypeptides as vaccines.

Some of the class of the RNA viruses now called retroviruses cause tumors. This was first recognized by Peyton Rous in 1911, when a retrovirus was found to elicit sarcomas (tumors of connective tissue) in chickens. But it was not until the mid-1960s that a large effort was made to understand the RNA tumor viruses found in many vertebrates, including mammals. Although no human RNA tumor virus had been identified when the work began, the assumption was that the studies would lead to an understanding of human cancer.

Retrovirus particles contain a reverse transcriptase that is encoded in the viral genome. After the virus infects a cell, the enzyme is activated and copies the single-stranded RNA genome into a double-helical DNA (Figure 10.5). Thereafter, the DNA is inserted by recombination into the cellular genome. New viral genomes are made when the cell's RNA polymerase transcribes the viral DNA into viral RNA, just as genes are transcribed in the cell's chromosome. The viral RNA also serves as a messenger RNA for synthesis of viral proteins. Viral coat proteins, reverse transcriptase, and viral RNA are then assembled into particles. The particles are extruded into the surrounding environment without destroying the cell; as they bud off the cell they acquire membranes in which are embedded molecules of another viral protein—the envelope protein.

Many of the cancer-causing (oncogenic) retroviral genomes studied in laboratory rodents carry coding segments other than the usual viral genes.

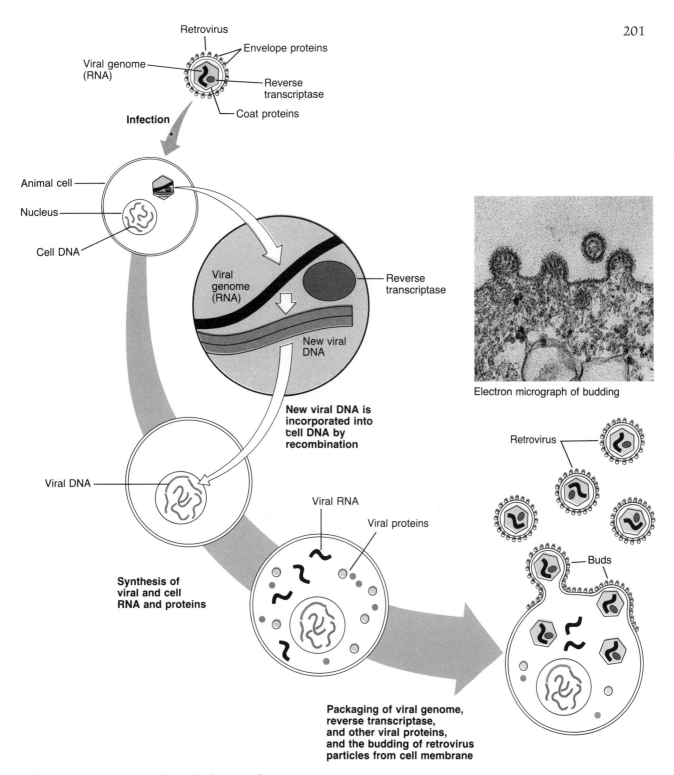

Figure 10.5 Events in the multiplication of a retrovirus.

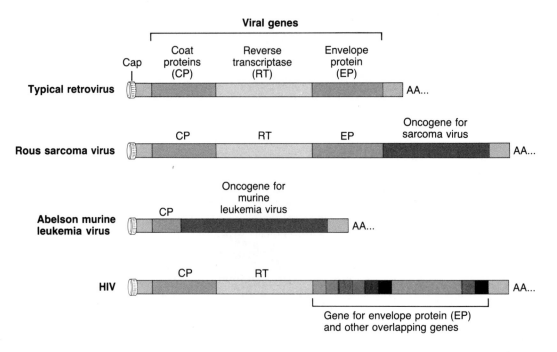

Figure 10.6 The RNA genomes of several representative retroviruses. The Abelson murine leukemia virus is unable to replicate without the help of another virus that can supply the missing RT and EP proteins. During infection with HIV, the gene for the envelope protein and other overlapping genes are expressed from alternately spliced mRNAs.

These segments are actually derived from cellular genes, but, when they are associated with the virus genome, they are called **viral oncogenes**. Figure 10.6 shows where viral oncogenes are located in some retroviral genomes. The first line is a typical, unaltered, retroviral genome. It contains genes for the virus's coat proteins, for reverse transcriptase, and for the envelope protein. The genome on the second line has an inserted oncogene. A different oncogene is in the genome shown on the third line; note that, in this virus's genome, the oncogene replaces the genes for reverse transcriptase and the envelope protein, so that the virus is defective and unable to reproduce by itself. However, defective virus genomes of this type can be propagated if the cells are also infected with a nondefective virus that provides the reverse transcriptase and envelope proteins. The genome of the AIDS virus, HIV, is on the last line; HIV is the most complex retrovirus known because it contains at least six genes in addition to the three found in all retroviruses.

Viral oncogenes account for the tumor-causing properties of many retroviruses. Each of the several classes of viral oncogenes encodes proteins that are capable of transforming normal cells into tumor cells. Generally, this means that the normal growth-controlling mechanisms fail, and the cell grows and divides unusually rapidly. Tumors may be solid—as for example, in sarcomas (connective-tissue tumors) and lymphomas (lymph-node tumors)—or they may consist of individual circulating cells, as in leukemias. Most tumors are actually clones of cells, all descended from a cell that underwent the initial event that resulted in transformation to a tumor cell.

How does a normal cellular gene become a part of a retroviral genome, and how does that make the virus oncogenic? Most likely, a retroviral genome acquires a cellular DNA sequence by recombination during an infection. The virus produced after such a recombination event will now have that cellular gene as part of its own genome. Following reverse transcription, the viral genome, with its newly acquired cellular gene, is incorporated into the DNA of newly infected cells. Thus, the tumor-causing retroviruses are actually transferring genes from one cell to another cell; this makes them analogous to transducing phages (p. 84). Usually, the viral oncogene is altered, compared with its normal cellular counterpart, by mutations that occur in the course of being acquired and propagated as part of the retrovirus. Following the insertion of such modified viral genomes into the DNA of the host cell, the expression of the concomitantly acquired cellular gene is regulated by the sequences that govern the transcription of viral genes, which are different from those associated with the normal cellular gene. In complex ways that are not yet well understood, the abnormal regulation or the abnormal product—or both—of viral oncogene expression causes the tumor. Just as cancer itself is not a single disease but a disparate set of disorders, so too different oncogenes encode different kinds of functions, and their capacity to promote tumor formation therefore depends on different mechanisms.

Cellular Oncogenes

WHAT DO the cellular genes that give rise to viral oncogenes do in normal cells? Most of them encode proteins that are involved in the control of normal cell growth. For example, some encode proteins that regulate transcription. Others encode growth factors that either stimulate or are required for cell growth. Still others encode cell-surface receptors that take up growth factors. Even yeasts and fruit flies have analogues of the genes that can become oncogenes in vertebrates. In these organisms, too, the analogous genes often have essential functions. For example, two yeast genes are related to a human cellular gene that, when mutated, contributes to the formation of human bladder and colon cancers. These yeast genes, known as *ras* genes, encode proteins that are needed to transmit signals from the extracellular environment to the interior of the cell. Yeast cells cannot grow if both *ras* genes are destroyed. However, if a normal human *ras* gene is inserted into a mutant yeast cell lacking both of its *ras* genes, the yeast cell acquires the ability to grow. Interestingly, when the cancer-causing form of the human *ras* gene is inserted into such mutant yeast cells, their normal growth control is also disrupted. This amazing finding, that human genes can supply essential functions in yeast, is also true for some other genes. Apparently, related gene products in organisms as different as yeasts and humans are functionally interchangeable. These observations are compelling evidence for evolution from a common ancestor. Thus, the cellular genes that give rise to viral oncogenes encode functions essential for life and were conserved throughout the evolution of eukaryotes.

Why don't the cellular genes that become viral oncogenes produce cancer cells under normal circumstances? Occasionally, they do become oncogenic, even without becoming associated with a retrovirus; but this occurs only if they are modified in certain ways. Thus, mutations in the coding sequence, or mutations that cause an increase in the level of expression, can change a normal gene into an oncogene. For example, if the normal human *ras* gene undergoes a mutation that alters one of several possible amino acids in the *ras* protein, the cell in which it occurs will become a tumor cell. Such altered *ras* genes are found frequently in the cells of bladder and colon carcinomas and occasionally in lung cancers. Rearrangements and chance duplications of some cellular genes can also alter their encoded products or their regulation, thereby creating a tumor cell. These studies emphasize how the discovery that retroviruses cause tumors in chickens eventually led to an understanding of how mutations in certain normal cellular genes result in the development of human tumors.

Another extension of this work is the discovery that there are also **tumor suppressor genes** in mammalian (including human) genomes. Proteins encoded by such genes can prevent cells from becoming tumor cells. Thus, a rare childhood tumor of the eye, retinoblastoma, is associated with mutations in both alleles of the "retinoblastoma" gene. A single good allele is enough to provide sufficient normal function and to prevent the appearance of tumors. In an unknown manner, the protein encoded by the retinoblastoma gene appears to drastically reduce the likelihood that a normal cell will become cancerous.

There has been a direct, if unpredictable, track from fundamental research on the RNA viruses that cause tumors in birds and nonhuman mammals to such urgent clinical problems as human cancer and the acquired immune deficiency syndrome, AIDS. AIDS is caused by a retrovirus called HIV that infects and kills a class of T lymphocytes, causing a devastating destruction of the immune response. Other human viruses, HTLV-I and HTLV-II, also infect T cells, but they cause lymphomas and leukemias. HTLV-I was, in fact, the first human retrovirus to be identified. Because of the prior intensive study of vertebrate retroviruses and the availability of recombinant DNA techniques, the structures of the HTLVs and of HIV (Figure 10.6) and their biological properties were elucidated very rapidly. This permitted the prompt development of cloned probes and antibodies as diagnostic tools. These tools allow the screening of blood supplies and the exclusion of contaminated blood from use in transfusions, an important element in controlling the spread of the diseases caused by these viruses. The detailed knowledge of HIV also suggests rational experimental approaches to the design of therapeutic procedures for AIDS.

The study of retroviruses has also had important ramifications for understanding the oncogenicity of some DNA viruses. For example, the tumor-producing potential of SV40 and of some papillomaviruses, adenoviruses, and herpesviruses after infection of newborn rodents is also associated with one or more viral genes. However, as far as we know, the oncogenes of DNA viruses have no normal cellular homologues. They appear to be required for the virus to multiply in their respective host cells, but, under certain circumstances, their expression can upset the normal cellular growth controls and cause the cells to become cancerous. By comparing the gene products of oncogenes found in oncogenic DNA viruses, retroviruses, the mutated cellular oncogenes and tumor suppressor genes, it might be possible to understand the many different ways in which tumors can arise.

Another important consequence of the study of retroviruses is their adaptation as recombinant vectors. These vectors have proven to be particularly useful for the introduction of new genes into fertilized mammalian eggs and, thus, into essentially all the cells of complete experimental animals, including germ-line cells; the resulting animals are called **transgenic.** The importance of transgenic animals to the study of gene expression and oncogenes is described in Chapter 12.

CHAPTER 11

Understanding Biological Systems

B IOCHEMISTRY and genetics, the antecedents of molecular genetics, epitomize what is called reductionist biology: Single biological elements—for example, a protein, an enzyme, or a gene—are examined in great detail through a series of narrowly defined questions and experimental approaches. This strategy can illuminate many characteristics of complex systems, including whole organisms, if an observable mutational alteration in a particular trait is the direct consequence of a single gene's malfunction. The virtue of such simple systems is illustrated by the red–green colorblindness common in human males. But mastering the intricacies of single genes, or even of a group of closely related genes, is only a first step in solving the higher-order complexities of living organisms. Unlike colorblindness, the vast majority of the characteristics of organisms arise from complex interactions between the products of many genes. To understand complex characteristics, one must study the properties of such cellular structures as ribosomes, chloroplasts, and mitochondria, of whole tissues, of physiological systems (such as the nervous system), of whole organisms, of populations of organisms, and of the interactions of organisms with their environments. The traditional alternative to reductionist biology—holistic biology—is based on the idea that the reductionist approach cannot provide a full understanding of organismal behavior and function, and that such an understanding can be achieved only if whole organisms are studied.

The expanding resources of molecular genetics are, in fact, applicable to the analysis of complex systems, including whole organisms. Complex properties can be described, and even manipulated, at the molecular level. Examples mentioned earlier include the regulation of serum cholesterol levels through a specific receptor on cell membranes and the formation of vertebrate immune proteins. Another example whose molecular features are becoming discernible is the formation and assembly of ribosomes, which is described later in this chapter. In this chapter, we also describe how some sets of genes are regulated in a coordinated manner during development, and we explain several different mechanisms whereby cells and tissues modulate the expression of particular genes. In part, this will summarize and review some points made in earlier chapters, but it will also introduce several new concepts.

Colorblindness: A Single-Gene Mutation

WHEN light strikes the retina, the absorbed energy is transformed into an electrical signal. After more than 40 years, we now have a detailed picture of how the absorbed light triggers a cascade of reactions that transmits the excitation to nerve cells. Two types of cells in the retina are responsible for absorbing the incident light. The **cone** cells of vertebrate eyes are responsible for the organism's color vision. Other, more abundant cells, the **rod** cells, give monochromatic perception and sensitivity to dim light. The pigment responsible for sensing light in rod cells is a protein called **rhodopsin**. Using the large amounts of pure rhodopsin that can be isolated from the eyes of cattle, it was possible to demonstrate that it is made up of two parts: a protein part called **opsin** and a light-absorbing small molecule that is a derivative of vitamin A. The bovine gene for opsin has been cloned, characterized, and used as a probe for the isolation of the human opsin gene. The two genes are similar enough in sequence to anneal together to form double-helical DNA.

Until recently, less was known about the color-sensitive components of cone cells. Some basic ideas about color perception had been deduced as early as the late eighteenth century by John Dalton, who was himself colorblind. The association of genes determining red–green colorblindness with the X chromosome was established as early as 1911 because of the prevalence of red–green colorblindness in males. But the paucity of cone cells made isolation of their light-receptor molecules (often called visual pigments) problematic, and, therefore, it was difficult to learn the molecular basis of color vision. It was learned that different color sensations were related not to the derivative of vitamin A that is the primary light absorber but rather to the proteins with which it is associated. By exploiting the weak similarity between the DNA sequences of the human opsin genes and the genes encoding the protein portions of the color photoreceptors, it was possible to clone the human genes encoding the proteins responsible for color vision.

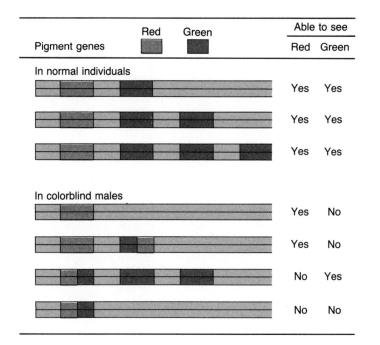

Figure 11.1 The multigene family that encodes the pigments for visual perception of red and green in normal and colorblind individuals.

The genes for the rod opsin (human chromosome 3) and the three cone opsins (blue on chromosome 7, red and green on the X chromosome, as seen on Figure 7.3, p.143) constitute a multigene family, apparently derived from a single ancestral gene by duplication and mutation of coding and transcriptional regulatory sequences. The amino acid sequences encoded by the genes for the red- and green-sensitive opsins are, for example, 96 percent identical. With the availability of the genes encoding the cone opsins, the stage was set for learning the mechanism of light absorption and signal transmission in cone cells. Another interesting question that can now be investigated is how the expression of the various opsin genes is regulated so that individual neighboring cells produce only the characteristic opsin in rod cells or one of the three different opsins in the cone cells responsible for detecting either red, green, or blue light.

The cloning of the genes has also provided an explanation for the high frequency of human colorblindness. In normal individuals, a single gene for the red-sensitive pigment and varying numbers (one to three) of genes for the green-sensitive pigment are arranged alongside each other on the long arm of the X chromosome (Figure 11.1). The analysis of DNA from males with different kinds of red–green colorblindness shows that their aberrant color perception is often associated with an altered number and arrangement of the genes encoding the color-sensitive pigments. Thus, those who lack the gene encoding the opsin that responds to green light can detect red colors but not green light. Similarly, those who lack (or have a damaged gene encoding) the opsin for responding to red light will not see red. A person in whom both types of genes are lost will fail to see both red and green. Probably, the similarity in DNA sequences and the close proximity of the different opsin genes makes them susceptible to recombinations between homologous chromosomes or between the different but homologous opsin sequences on the same chromosome. Such recombinations can lead to deletions (Figure 8.4, p.164) and other rear-rangements, as has already been noted for the gene encoding the cell-surface protein that clears cholesterol from the blood (Figure 8.8, p. 169).

Coordinating Many Genes

RIBOSOMES are essential to all cellular functions because they are required for protein synthesis. Because the production of ribosomes requires many different genes encoding three or more different RNAs and about 100 proteins, it is a particularly good example of the complexities that arise in trying to learn how separate genetic systems—ones dependent on synthesis of different kinds of RNA and many types of proteins, and their orderly assembly into a functional unit—are coordinated in time and cellular location. We know, for example, that, shortly after cell division, the transcription of rRNA genes triggers the formation of a nuclear body, the nucleolus (Chapter 1). Newly made ribosomal proteins and several enzymes must be transported from the cytoplasm into the nucleolus, and several additional RNAs must also enter the nucleolus. Recall also that, while the transcription of genes encoding ribosomal proteins is carried out by one type of RNA polymerase, the formation of the rRNAs requires two other types of RNA polymerase. Therefore, the outputs of three different independent transcription systems must be coordinated with each other and with the translation of the ribosomal and other proteins. Finally, all of the components of the ribosome must be assembled in the nucleolus, and the finished product—the ribosome—must be exported outside of the nucleus before it is able to function in protein assembly. Learning how this complex process occurs is a formidable undertaking, but it is the kind of problem in which the new tools of molecular genetics should prove to be indispensable. In fact, these tools have already clarified some features of this complex machinery and its regulation.

Turning Genes On and Off

VIRTUALLY every cell in a complex organism contains the same complement of DNA—the cellular genome. Some genes actively produce their gene products in all cells. For example, ribosomes occur in all cells, and genes for rRNA and ribosomal proteins are expressed in all cells. Such genes are often called "housekeeping" genes, because they are involved in the basic functions common to all living cells. However, many genes are active only in specific cells or tissues. The genes encoding the opsins are a case in point. As already mentioned, they are expressed only in the cells of the retina, and, even there, rod cells express one opsin gene and cone cells each express only one of the several other opsin genes needed for color perception (that is, the green-, red-, or blue-sensitive pigments). Yet the thousands of different cell types and the many specialized tissues that they form all arise from a single cell, the fertilized egg cell. Differential gene expression imparts to these cell types their unique shapes, functions, and capabilities. Through the successive cell divisions, starting with a fertilized egg, cells arise in which certain genes are turned on and others are turned off. Moreover, these on–off controls are activated at very precise times during development. In addition, the position of a cell in the developing embryo influences which genes are turned on or off. Understanding these extraordinary events is one of the most challenging and interesting problems in biology.

How are the positional and temporal regulation of cell- and tissue-specific gene expression achieved? Primarily by controlling the initiation of transcription. As described in Chapter 6, this requires interaction between specific DNA sequences in the vicinity of a gene and specific proteins. The proteins may activate or inhibit gene transcription. While the transcriptional on–off switches are probably not absolute, experimental evidence suggests that rates of gene expression may be regulated over a millionfold range.

Molecular genetic and biochemical studies are revealing the detailed mechanisms governing the differential regulation of gene expression. The regulatory signals in DNA consist of complex arrays of relatively short DNA sequence motifs. Each motif is a binding site for a specific protein, a **transcription factor**. For example, genes that are uniquely transcribed in white blood cells, such as antibody genes, contain an array of sequence motifs that are bound by transcription factors, some of which are restricted to white cells. Similarly, genes expressed only at certain times, or only under particular environmental conditions, contain sequence motifs that interact with proteins that are present or active only at those times, or only under those conditions. Turning a gene on or off depends on the particular assortment and arrangement of DNA sequence motifs, the availability of the corresponding transcription factors, and the way the factors influence the transcription machinery itself. We still do not know if binding of the multiple transcription factors to the gene's regulatory regions facilitates the assembly of the transcription machinery, or the operation of the RNA polymerase, or both.

An important corollary of this general mechanism is that different cell types have different active regulatory proteins. But this explanation doesn't go far enough, because we now must explain what controls the presence or absence of the regulatory proteins. Moreover, there are hints that more than specific regulatory proteins are at work here. It is possible that relatively large chromosomal regions can be "opened" or "closed" for gene expression by changes in chromatin structure. In at least some instances, transient modifications of a gene's DNA sequences by the addition of methyl groups to cytosines in the vicinity of the promoter modulates gene expression. Understanding these matters fully will require a great deal more research.

An extraordinary feature of the regulation of gene transcription by a combination of DNA sequence motifs and corresponding protein transcription factors is its universality. Notably, corresponding transcription factors from such diverse sources as yeasts, flies, and mammals are interchangeable. A yeast or fly transcription factor can interact with the corresponding mammalian DNA sequence motif and with the other mammalian proteins required to turn on the mammalian gene. Similarly, some mammalian factors can replace the corresponding factor in yeast cells. This universality implies a remarkable conservation of structure in the course of evolution.

Besides control of the initiation of transcription, there are other specialized mechanisms that regulate gene expression. Among the more unusual ones are those that, like the formation of functional immune-protein genes in vertebrates, depend on the rearrangement of genomic DNA. Polypeptide levels can also be modulated by controlling the rate of mRNA maturation from the primary transcript, or by controlling the mRNA's stability, or by regulating the way in which the polypeptides are modified, transported, or protected against destruction.

With the availability of recombinant DNA methods, we are no longer dependent on chance mutations to understand normal gene function. Specific types of mutations in selected genes can be designed and introduced into cells or into the early embryos of certain experimental organisms. Here, we describe a few examples from the broad range of current studies on the control of gene expression in already differentiated cells. Work on embryonic and adult organisms is described in Chapter 12.

Differential gene expression in specific cells and tissues includes not just single genes, but sets of genes whose expression is turned on and off in a coordinated manner. The various members of some multigene families encode related or identical proteins that are expressed in different tissues or at different times during development. In the mouse, for example, two genes encode the related proteins α-fetoprotein and serum albumin. Both proteins are components of blood serum. The two genes are next to each other on one mouse chromosome, and both proteins are made in embryos. Shortly after birth, expression of the α-fetoprotein gene decreases to as little as one ten-thousandth of its previous level; thereafter, serum albumin, but not α-fetoprotein, is found in the adult mouse's blood. The change is regulated by a protein encoded by another mouse gene, but the way in which this regulatory gene is itself regulated and how it turns off the α-fetoprotein gene after birth is still a mystery.

The expression of the genes for the α- and β-polypeptides of hemoglobin is restricted to vertebrate red blood cells. At different developmental stages, different members of the groups of α- and β-polypeptide genes are expressed. To understand how this regulation is achieved, current work is focusing on the DNA sequences and regulatory proteins responsible for turning the embryonic, fetal, and adult α- and β-globin genes on and off at the proper times. Such studies are likely to inform us about why, in some instances (such as when the β-polypeptide gene is deleted), the expression of the fetal gene continues throughout adult life. Does this suggest the existence of interconnected regulatory systems?

Fortuitously identified mutations also contribute to an understanding of the expression of the hemoglobin genes. For example, the human genetic diseases known as thalassemias are caused by a failure to synthesize adequate amounts of either the α- or the β-polypeptides of hemoglobin. In several thalassemias, mutations that create stop codons in the middle of the coding sequence prevent completion of the translation of mRNA. Other kinds of mutations prevent normal intron splicing and therefore result in untranslatable mRNAs. Still others result from the complete loss of one or more of the α- or β-polypeptide genes.

The pancreas, which appears superficially to be a single organ, is in reality a conglomerate of several different cell types. This is also true of many other organs. Each pancreatic cell type synthesizes and secretes highly specialized proteins. Insulin synthesis occurs in only one cell type, found in the so-called pancreatic islets. This hormone is then carried throughout the body by the blood stream. Other pancreatic cells synthesize digestive enzymes, such as chymotrypsin, trypsin, and amylase, and these are deliv-

Figure 11.2 (*facing page*)   A typical mammalian insulin gene showing its intron, coding sequences (exons), and the regulatory sequences that are required for transcription in the islet cells of the pancreas. The gene transcript (RNA), the polypeptide translated from the messenger RNA (preproinsulin), the partially processed polypeptide (proinsulin), and the final active insulin molecule are shown below.

ered to the digestive tract. Nevertheless, all the genes for these proteins are present in all cells. And yet, they are expressed only in particular pancreas cells. What processes and elements are responsible for this exquisitely selective gene expression? And how, during development, are these precise and limited synthetic capacities acquired?

Near the beginning of the insulin gene, in the 5′ flanking region, there are several sequence elements that are critical for ensuring proper regulation of insulin production (Figure 11.2). The binding of different regulatory proteins to these sequences modulates the transcription of the insulin gene. In cells that do not synthesize insulin, there are proteins that bind to this region and thereby prevent the transcription of the gene. In pancreatic islet cells, however, there are proteins that activate transcription of the insulin gene when they bind to this region. Similarly, close to the genes encoding the digestive enzymes, there are specific DNA sequences that bind cell-specific proteins and influence the synthesis of the digestive enzymes in the other pancreatic cells. Thus, during differentiation, the characteristic synthetic capacities of the two cell types must be established by the switching on (and off) of specific regulatory proteins. The details of the switching remain to be worked out, but the general strategy is clear and is duplicated many times over in various organs.

Hormones and Other Polypeptide Messengers

STILL other controls on the rate of insulin synthesis can occur after translation. The polypeptide translated from insulin mRNA (preproinsulin) is larger than the active hormone (see Figure 11.2). Cleavage by special enzymes removes a short piece from the amino end to form proinsulin; then the central part of the polypeptide chain is eliminated, leaving two short polypeptide chains that comprise insulin. These are held together by chemical bonds formed between sulfur atoms in the cysteine amino acids in each chain. The two polypeptides together constitute the active hormone. Such processing of inactive protein precursors plays an important role in modulating the concentrations of the active forms of many hormones.

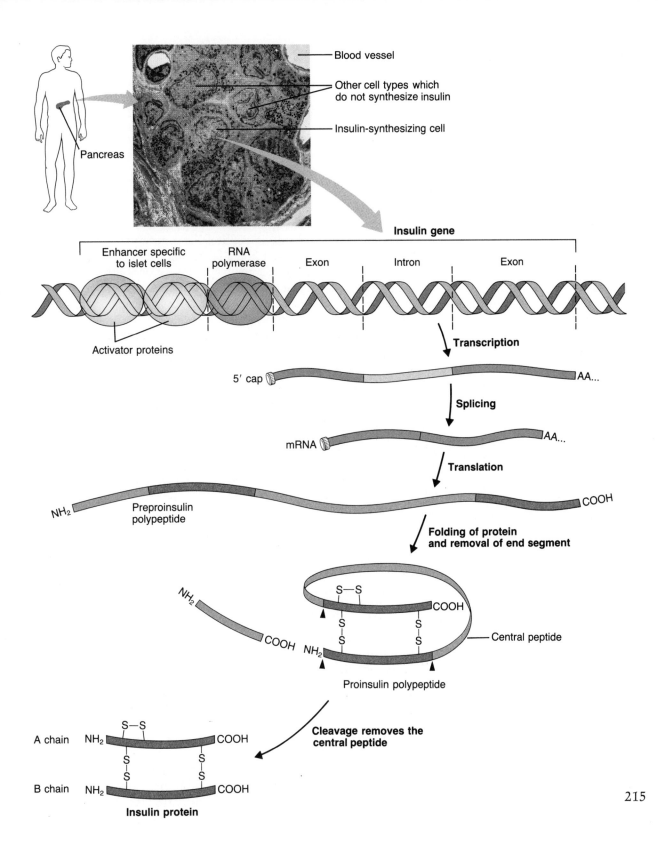

Blood vessel

Other cell types which do not synthesize insulin

Insulin-synthesizing cell

Pancreas

Insulin gene

Enhancer specific to islet cells

RNA polymerase

Exon

Intron

Exon

Activator proteins

Transcription

5' cap

AA...

Splicing

mRNA

AA...

Translation

NH₂

Preproinsulin polypeptide

COOH

Folding of protein and removal of end segment

S—S

NH₂

COOH

S S

S S

COOH NH₂

Central peptide

Proinsulin polypeptide

Cleavage removes the central peptide

A chain

NH₂

S—S

COOH

S S

S S

B chain

NH₂

COOH

Insulin protein

215

The levels of an active protein are frequently regulated after its synthesis. This can occur whenever the newly synthesized protein requires chemical modification in order to be in an active state. Such modifications are especially common among the myriad polypeptides that serve as messengers between different cells in multicellular organisms. Hormones such as insulin, for example, circulate in the blood and coordinate the activities of widely separated cells. Other polypeptide hormones act over shorter distances to influence the activity of cells in the vicinity of the secreting cell. For example, certain short polypeptides, such as the endorphins, serve as signals or messengers between nerve and other cells.

Basically, all **peptide messengers** work in a similar fashion, regardless of whether they are wide-ranging, like insulin, or confined to small spaces, like the neurotransmitters. In the initial step, they recognize and bind to highly specific protein receptors on the external surfaces of particular target cells. This interaction then triggers a precise response in the cell's interior, the nature of which depends on the properties of the receptive cell: These include such complex responses as cell growth, secretion of another polypeptide, expression of a particular gene, firing of a neuron, initiation of specific fixed behavioral patterns, and so forth. Note that this phenomenon resembles the way in which surface receptors in cells of the immune system influence intracellular events when they bind to antigens. For the present, however, we do not understand fully how the binding of peptides or other molecules to receptors at the cell surface activates intracellular responses and even specific gene transcription.

Although many different peptide messengers are encountered by all cells, each cell type can only respond to those messengers for which it has specific receptors. Any single peptide messenger can, however, elicit different types of responses, if receptors for that messenger occur on different cell types. This is similar to what happens when a salesman rings many doorbells. The person ringing the doorbells is the same each time, but the response that is elicited will vary from house to house. The peptide messenger cholecystokinin is a good example. Cholecystokinin is secreted by certain intestinal cells and travels through the bloodstream. Many cells are completely by-passed, but, in the gall bladder and in the brain, cholecystokinin meets cells with matching receptors. Cells in the gall bladder respond by increasing the flow of bile into the intestines. In the brain, however, cholecystokinin acts as a neurotransmitter.

Many different intercellular peptide messengers are known. These, together with other intercellular messengers that are not polypeptides (for example, such small molecules as acetylcholine, epinephrine, norepinephrine, and the prostaglandins), coordinate the physiological activities of, for example, the nervous and circulatory and digestive systems.

The genes that encode the many types of polypeptide intercellular messengers generally encode proteins that are several hundred amino acids long. However, the polypeptide messenger itself may be as short as five amino acids. For example, the two enkephalins that act as natural opiates in the brain and regulate contractions in the gut have the following amino acid sequences:

Tyr-Gly-Gly-Phe-Met   (met-enkephalin)

Tyr-Gly-Gly-Phe-Leu   (leu-enkephalin)

The very long polypeptides initially produced by expression of the enkephalin genes are shortened in a complex series of events that take place within the intracellular bodies that secrete the enkephalins. Frequently, additional chemical modifications are made to the polypeptides. Finally, the precursor is trimmed down to yield one or more active polypeptide messengers. For example, a polypeptide 263 amino acids long that is encoded by a single gene in mammals produces six molecules of met-enkephalin and one of leu-enkephalin.

The characteristic properties of mammalian genes encoding intercellular messengers originated early in evolution. The same general strategies are used to encode and synthesize the small secreted polypeptides, called pheromones, that permit individual yeast cells to communicate their mating types to one another. Small polypeptides that elicit and coordinate fixed behavioral patterns in invertebrates, such as marine slugs, are produced in a similar way. In the slug, as in mammals, a set of polypeptide messengers permits many different cell types in the animal to communicate with each other, so that the animal can coordinate its behavior.

DNA Rearrangement Can Regulate Gene Expression

DIFFERENTIAL gene expression that depends on DNA rearrangement is widespread in nature, although in any particular organism it appears to operate on only a very small number of genes. One such programmed reorganization, described in Chapter 8, is the assembly of genes encoding the immune proteins. It depends on DNA sequences that act as recombination sites and on proteins that carry out the recombinations. These particular proteins are encoded in the genome and are expressed in T and B cells. The rearrangements have three functional consequences: First, they lead to the formation of complete genes by joining DNA segments that are not contiguous in the germ-cell DNA. Second, they allow the efficient initiation of transcription by bringing the required regulatory elements close together. Third, they generate diversity in the DNA segments encoding the immune proteins and thereby provide for an enormous repertoire of antigenic responses. Keep in mind that, because the rearrangements do not occur in germ cells, each generation begins life with the same potential for its immune system.

Another example of gene regulation by DNA rearrangement is the means by which trypanosomes, the cause of sleeping sickness in humans and related diseases in cattle, avoid the immune system of the infected animal. These small protozoans are covered by many molecules of a surface protein. There are as many as a hundred slightly different genes in the trypanosome's DNA, each of which encodes a slightly different variant of the surface protein. By switching sequentially from one surface-protein gene to another, some individual trypanosomes manage continuously to escape immune destruction. In a sense, the trypanosome dons new clothes to escape detection.

Prokaryotes also use DNA rearrangements to alter gene expression. *Salmonella* bacteria are able to move (swim) in a liquid environment because of hairlike flagella that extend from their cell membranes. These flagella come in two forms, depending on which of two proteins is used in their construction. The orientation of one particular DNA segment, 1000 base pairs long, determines whether the gene for one or the other protein is active. In one orientation, the DNA segment encodes one form of the flagellar protein, and in the other orientation, the other flagellar protein is made. This enables the bacteria to evade the antibody response made by an infected individual against one of the forms of flagellar protein. A bacterial enzyme acting at particular DNA sequences at either end of the segment flips the DNA segment around, each time breaking and rejoining the necessary bonds.

Yeast cells often contain only one of each pair of homologous chromosomes, like the germ cells of more complex eukaryotes. Such haploid yeast cells multiply by budding off daughter cells. However, two haploid cells can fuse to form diploids as long as the two cells are of opposite mating types (or sexes). Many haploid yeast cells switch back and forth

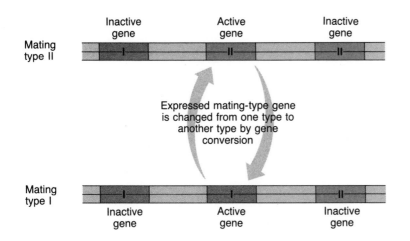

Figure 11.3 Mating-type switching in yeast. The yeast's mating type is determined by which of two genes (I or II) is in the active position. Switching occurs when the gene in that position is changed to the other type by gene conversion (Figure 8.2, p. 163).

between sexes. This depends on which of a pair of mating-type genes is present in a special chromosomal DNA site—the so-called mating-type position (Figure 11.3). Virtually every time the yeast cell buds, it switches sexes by replacing the gene in the active mating-type position with a copy of the alternate gene. The process is akin to placing one of two different tapes (the gene copies) into a cassette player (the mating-type position). In that analogy, either one of two recorded messages will be played, depending upon which tape is in the player. Remember that cassette tapes are themselves copies of master tapes. In yeast, the "master" copies of the mating-type genes are always stored intact at other chromosome positions.

Various other kinds of DNA rearrangements influence gene expression. Sometimes, for example, multiple copies of a gene are made at special times in development in order to meet a large demand for the encoded material. Altogether, there are enough examples to lead to the conclusion that it is not precisely true to say, as we have, that every cell in a complex organism has the same DNA as the fertilized egg from which it arose. Nevertheless, the overwhelming majority of genes and DNA segments that have been cloned to date have the same structure in germ cells and in body cells.

12

Manipulating Biological Systems

T HE POTENTIAL of molecular genetics extends beyond increasing
our ability to understand biological systems. It also includes a novel
competence: to alter individual genes, cells, and whole organisms in
a precise manner. The ability to make such modifications is not an entirely
new enterprise. For millennia, humans have successfully modified the genetic
makeup of organisms through selective breeding. All of our agriculturally
productive plants and animals are the result of thousands of years of carefully
manipulated breeding. Contemporary horses, cows, dogs, wheat, corn,
grapes, and tomatoes are not the natural products of evolution but rather
the results of human intervention in genetic processes. Debilitated viruses,
selected for their altered properties, are the basis of successful vaccines for
polio and other viral diseases. Microorganisms have been manipulated to
increase the yield and potency of the antibiotics they produce. And it is
not only humans that affect the genetic traits of other species. Organisms
of all types influence the genomes of other species, including those of
humans: subtly (for example, by competing for certain food supplies and
thereby applying selective pressure for traits that permit the use of alternate
nutrients) and boldly (by killing off large portions of populations that lack
the genetic makeup to protect themselves).

Still, there is no question that the new biology gives human beings
tools that are far more precise for manipulating both themselves and other
species. In one sense, this is advantageous, because the new methods reduce
the chance that undesirable and undetected modifications will accompany
the sought-after change, a serious drawback of the nonspecific breeding
techniques. But the influence over nature that derives from the new techniques is qualitatively different from what was previously possible. Inev-
itably, it raises political, social, and cultural issues in addition to scientific
questions. For this reason, the euphoria of most biologists is tempered by
awe and by an acceptance of the fact that society at large must participate
in decisions concerning the genetic modification of living organisms.

The concepts and techniques of molecular genetics will be used to modify biological molecules and systems in many presently unknown ways. Currently, the approaches range from the synthesis of altered proteins to the introduction of new genes into plants and animals, thereby endowing these species with new properties. The remainder of this chapter gives an overview of the current experimental systems.

Modifying Proteins

THE EARLIEST practical goal of recombinant DNA research was the production of medically and economically important proteins, such as vaccines and hormones (for example, insulin, growth hormone, and oxytocin), that are important tools of clinical medicine. The idea was to clone the gene encoding the polypeptide and insert it into a plasmid that replicates in the bacterium *E. coli*. The *E. coli* cell would then transcribe and translate the inserted gene and become an efficient factory for making the desired protein. By growing the *E. coli* in large production vats like those used to make beer, large quantities of the encoded protein could be synthesized. What appeared to be a relatively straightforward scheme turned out to be more complicated than expected. The problems encountered highlighted the differences between prokaryotic and eukaryotic genes. Circumventing the difficulties helped to achieve a better understanding of gene expression in both kinds of organisms.

One problem is that most eukaryotic genes have introns, whereas *E. coli* genes do not; the bacteria lack splicing mechanisms and thus are unable to produce the correct mRNA from a eukaryotic gene. Further, the translation products of many eukaryotic genes, such as the polypeptide hormone precursors described in Chapter 11 (pp. 214–217) require additional processing to become active proteins. Unfortunately, *E. coli* cells do not carry out these processes. Finally, success in obtaining good yields of many eukaryotic proteins from *E. coli* is thwarted by the proteins' toxicity to the bacterial cells, or by their degradation by bacterial enzymes, or by their tendency to aggregate into particles inside the bacterial cell.

The solution to the intron problem is to use a cloned cDNA rather than a cloned gene because the cDNA, being a copy of mRNA, lacks the introns that were removed by splicing. The problem of appropriate processing of the initial translation product can be circumvented by having *E.*

coli produce the active polypeptide directly. A DNA insert encoding only the desired final product is inserted into the recombinant DNA plasmid rather than the cDNA corresponding to the natural eukaryotic mRNA. Insulin, as we have seen, is composed of two different polypeptide chains, the A and B chains (Figure 11.2). Although both polypeptide chains are the products of a single gene, the initial long polypeptide made by translating the insulin mRNA is cleaved several times to yield the A and B chains that constitute functional insulin. *E. coli* has been adapted to synthesize human insulin by using two different recombinant plasmids (Figure 12.1). One plasmid contains a DNA insert encoding the A chain, and the other plasmid encodes the B chain. Neither insert has any introns. The two plasmids are introduced into different *E. coli* populations, and clones producing either the A chain or the B chain are isolated. Each of the polypeptides is thus synthesized in a different population of *E. coli* cells. The two purified polypeptides are then mixed to form insulin. Because DNA segments encoding only the A or B chain could not be made by copying natural mRNA, they were synthesized chemically. Using the genetic code, it was straightforward to convert the known amino acid sequence of insulin into an appropriate nucleotide sequence for an mRNA.

The problem of the toxicity of foreign proteins to *E. coli* can be addressed by using *E. coli* promoters that can be induced to make mRNA encoding the foreign protein. Recall that, in Chapter 3 (p. 71), we mentioned that the sugar lactose can serve as an inducer for producing β-galactosidase protein by binding to the repressor, thereby activating the β-galactosidase gene promoter. In the absence of lactose, no β-galactosidase is produced. *E. coli* cells harboring a cloned eukaryotic or viral gene under the control of the β-galactosidase gene promoter can be grown to a high population density in the absence of lactose. Then the lactose is added to initiate transcription and translation of the cloned gene. In this way, formation of the toxic foreign protein is delayed until the cells are nearly fully grown, thereby preventing its interference with the process of cell multiplication. The amount of foreign protein produced can also be increased by using strains of *E. coli* deficient in enzymes that degrade proteins.

In some cases, it proves more useful to cause a recombinant gene to be expressed in yeast or animal cells rather than in *E. coli*. In a eukaryotic environment, appropriate modifications of the new polypeptide chain can occur (for example, protein cleavage or the addition of sugars or phosphates). Coding sequences can even be redesigned to use the cell's normal secretory pathways so that the polypeptide collects outside the cell. This trick depends

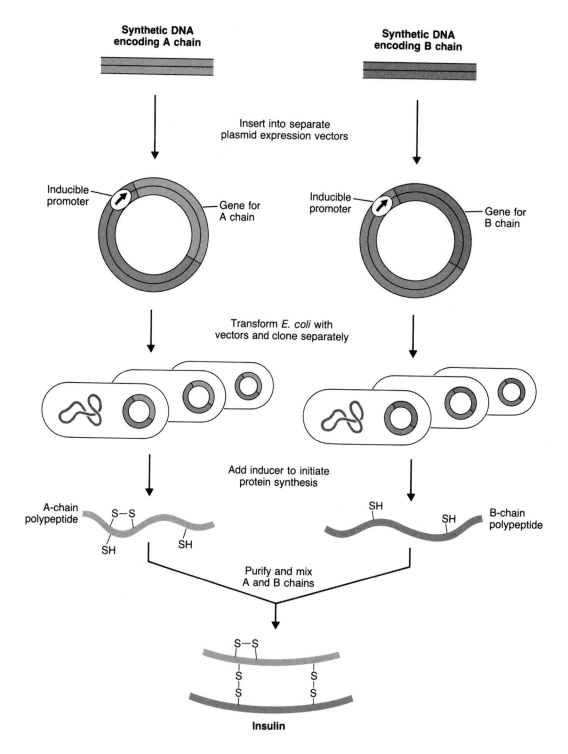

Figure 12.1 The production of human insulin in *E. coli*.

on a special short sequence of amino acids that serves as a "shipping tag." The tag is used by the cellular machinery to guide a protein through the cell membrane to the outside environment. Nucleotide sequences that encode the tag are occasionally part of a gene's coding sequence, but they can also be added to any gene's coding region.

Although it took longer than the optimistic predictions made in the mid-1970s, important polypeptides produced by recombinant DNA methods are now in use. The vaccine against hepatitis B virus has already been mentioned. Clinical trials of a vaccine against malaria, a disease that affects more than 100 million people each year, are now being carried out. Also under way are trials of vaccines potentially useful for controlling the spread of AIDS. Human insulin and growth hormone are now available. Lymphokines, the molecules that control the growth of white blood cells (lymphocytes), are being tested as therapeutic agents against cancer. Other proteins not previously available are proving to be clinically important. These include tissue plasminogen activator, a powerful agent for alleviating certain kinds of heart attacks, and erythropoietin, which is being used to treat the anemia associated with dialysis treatment of patients lacking functional kidneys. Still other polypeptides produced by recombinant DNA techniques are used in veterinary medicine. For example, pigs are being protected from disease by a vaccine against an important bacterial toxin.

Quite apart from its application to the synthesis of therapeutic polypeptides, the ability to synthesize substantial quantities of specific polypeptides has impressive implications for the study of protein structure and function. The three-dimensional structure—and consequently the biological function—of each protein depends on its unique sequence of amino acids. Modification of the side chains of individual amino acids (or of groups of amino acids) by chemical methods strikingly alters the ability of a protein to assume the correctly folded shape and consequently to carry out its precise normal function. Analysis of proteins produced by genes containing natural mutations in their coding regions tells the same story. The globin defect in sickle-cell anemia, which was described in Chapter 2, was among the earliest illustrations of this principle, and it remains one of the most dramatic.

Rather than depending on random mutations or nonspecific chemical modifications, a scientist can now make systematic changes in the amino acid sequence of a protein by changing the nucleotide sequence of its cloned gene or cDNA. A variety of techniques allow the substitution of one nucleotide by another in a DNA molecule. The altered DNA can then be cloned, yielding an artificially mutated gene that directs the synthesis of a mutant protein in an appropriate host cell. The physical, chemical, and functional properties of the mutant protein can then be studied. Regions of the protein that are important for forming the correctly folded structure can be delineated, and those amino acids that contribute to the catalytic properties of enzymes can be defined. It is possible to redesign enzymes and proteins to change the conditions for their optimal activity, or their affinity for the molecules on which they act. Antibodies are even being modified to act like enzymes. Thus, the ramifications of recombinant DNA techniques have been extended to chemistry and will foster an understanding of proteins at a previously impossible level of sophistication. A few examples will illustrate the potential of these investigations.

There is an inherited disease in which patients accumulate abnormal amounts of the polypeptide precursor of insulin (Figure 11.2, p. 215). The disease results from a mutation that changes the tenth amino acid from the amino terminus of the insulin B chain—histidine—to an aspartic acid. As a consequence of the mutation, the precursor escapes the processing reaction that would result in the removal of its central region to yield the A and B chains of insulin. By producing large quantities of the altered insulin precursor from its cloned DNA sequence, it was possible to study the properties of the mutant precursor. Besides being inefficiently processed to insulin, the precursor, as well as the corresponding altered insulin itself, binds more tightly to insulin receptors on cell surfaces than do their normal counterparts. Current experiments are aimed at understanding how the single amino acid change in the precursor molecule reduces the efficiency of normal processing. These investigations are also likely to tell us more about how proper insulin levels are maintained in normal individuals and to aid in the design and testing of rational therapies.

Trypsin is an enzyme that cleaves polypeptides into smaller chains. It is an important enzyme in food digestion and also a useful tool for the study of protein structure. Trypsin cleaves polypeptides by breaking the peptide bonds that connect amino acids. It is quite finicky, preferring certain peptide bonds over others, depending on the particular amino acids around the bonds. The amino acid in position 226 in the trypsin molecule helps to form a "pocket" in the folded enzyme that fits the region of the polypeptide to be cleaved. If the space within the pocket is decreased by replacing the normal amino acid at position 226 with a larger amino acid, there is a concomitant change in the kind of peptide bond that is preferentially cleaved by the enzyme. Experiments of this sort show how enzymes work, and they can lead to redesigning naturally occurring enzymes for innovative industrial and clinical purposes.

Genetic Engineering of Animals

PROTEIN chemists change the nucleotide sequences of cloned genes to produce pure altered proteins in bulk. Biologists are interested in studying the effect of alterations in protein structure, regulatory regions, or genomic organization on the properties of whole cells and organisms. In this way, they can better understand normal and abnormal physiological states. The introduction of altered genes into cells and whole organisms can achieve this aim, and the most profound implications of the recombinant DNA techniques lie precisely in their ability to accomplish such alterations. It is here that biology changes from its traditional concern with describing how living things are constructed and how they function into being a manipulative science that can make permanent, heritable changes in organisms. The term "genetic engineering" is apt. Moreover, it is here that the reductionism inherent in molecular genetics turns to the study of whole cells and organisms and the wide-ranging effects of single genes on physiological, anatomical, and developmental systems.

The transformation of bacteria or eukaryotic cells in culture by recombinant vectors containing genes, cDNAs, and altered versions of DNA or cDNA segments is one of the basic techniques in molecular genetics. When the cells in question are bacteria or yeast cells, then of course whole organisms are being manipulated, albeit single-celled organisms. When a correction of a mutation in a yeast cell or a bacterium is made experimentally, it can be described as **gene therapy**. In Chapter 10, we described a therapeutic effect of a human *ras* gene on "sick" yeast cells—sick because they lack their own *ras* genes. However, the extension of such directed modifications to gene therapy of multicellular sexually reproducing organisms requires conceptually and experimentally different approaches, as described in the following paragraphs.

One approach to the genetic modification of multicellular organisms is to change a gene in only one type of differentiated cell—that is, to modify only somatic (body) cells. The current experimental approach requires several steps: (1) removing cells from the organism; (2) placing them in a dish containing a nutrient medium in which they can survive, or grow and divide; (3) transforming them with a vector containing the gene of interest (or the corresponding cDNA); and (4) reintroducing the cells into the individual from which they were obtained. This is the basic protocol presently being considered for therapeutic intervention in some human genetic diseases. The experiments utilize mammalian bone-marrow cells, which include the progenitors of the circulating blood cells (Figure 12.2).

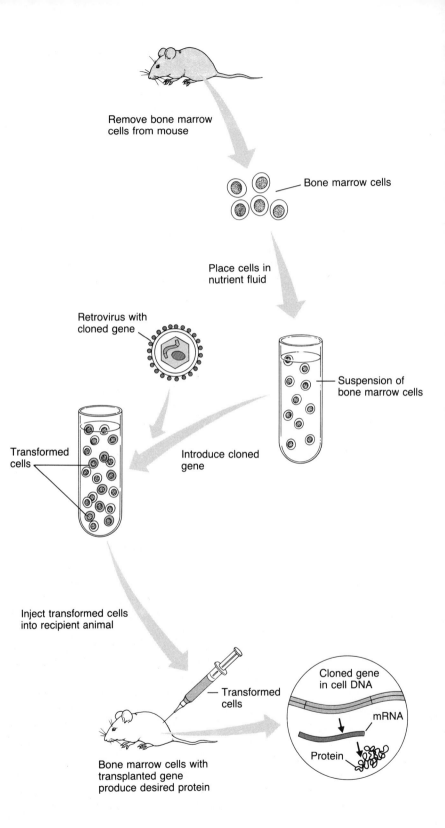

Remove bone marrow
cells from mouse

Bone marrow cells

Place cells in
nutrient fluid

Retrovirus with
cloned gene

Suspension of
bone marrow cells

Transformed
cells

Introduce cloned
gene

Inject transformed cells
into recipient animal

Transformed
cells

Cloned gene
in cell DNA

mRNA

Protein

Bone marrow cells with
transplanted gene
produce desired protein

Figure 12.2 (*facing page*)   Somatic-cell gene therapy: the steps in altering the genome of bone-marrow cells and reintroducing the cells into a whole mammal.

Transformation is being attempted in a variety of ways, mainly using vectors, derived from retrovirus genomes, that carry functional genes of special interest. In a particularly interesting trial, a child with a genetic disease that causes a severe deficiency in the functioning of the immune system is being treated. Humans suffering from this disease are homozygous for mutations that damage the gene encoding the enzyme adenosine deaminase. Thus, the therapy is attempting to transplant the normal gene encoding adenosine deaminase into the child's lymphocytes. There are several other diseases that might be ameliorated by introducing a functional gene into blood cells. However, it remains to be seen if the results of the current tests are promising enough to warrant additional therapeutic trials.

The alteration of somatic cells does not introduce heritable changes into multicellular organisms, because germ cells are sequestered very early in development. To introduce heritable changes, the germ cells themselves must be modified. So far, such modifications have been successful only in three types of organisms: fruit flies, experimental mammals, and plants.

In fruit flies, germ cells are modified by using a transposable element called a P element. A recombinant vector carrying a gene of interest inserted into a P element is injected into very early fly embryos. The P element, along with the embedded gene, gets transposed from the vector into genomic DNA (Figure 12.3). Adult flies that develop from such embryos frequently have the P element in their germ-cell DNA, and thus their offspring inherit and express the new gene.

Flies and other organisms carrying experimentally introduced genes that can be passed on to succeeding generations are called **transgenic**; the newly introduced gene is a **transgene**. This technique provides extraordinary opportunities for studying the intricacies of development and differentiation. One such opportunity relates to the poorly understood effect of neighboring DNA sequences on the control of gene transcription. Note that the position of the new gene—the transgene—in the transgenic fly's DNA will be different from the normal position of the corresponding gene in the genome. Moreover, in a series of independently prepared transgenic flies, the transposition process will have placed the transgene into different genomic locations—for instance, in different chromosomes. Among the questions that can then be investigated are: What regulatory DNA segments must accompany the transgene, if its function is to mimic that of the corresponding gene in the normal chromosomal position. That is, what DNA segments are required if the gene is to be expressed at the right time and in the right cells during development? Is the chromosomal

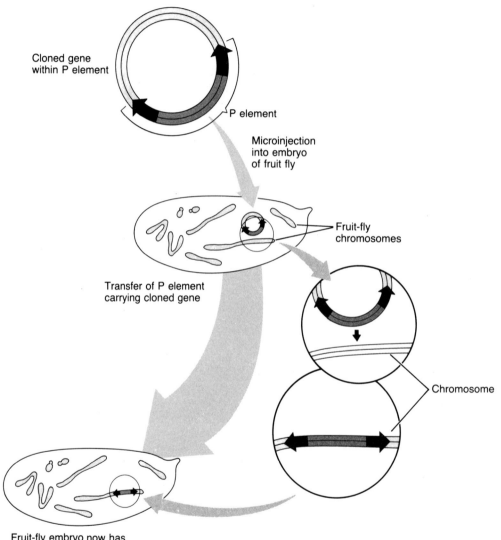

Figure 12.3 The germ-line modification of fruit flies by insertion of a P element carrying a fly gene. The gene in the P element is functional; the one in the recipient fly genome is a nonfunctional mutant.

position of the transgene important for its appropriate regulation in particular kinds of cells or at particular times during development?

These same questions motivate the introduction of foreign DNA sequences into the germ-line cells of experimental mammals. Transgenic mice
are the preferred experimental models for studying basic biological questions. Transgenic fish, sheep, cows, rabbits, and pigs are also studied, with
the eventual aim of improving the yields of important foods and materials
or of creating animal sources of useful therapeutic proteins.

Transgenic mice can be produced in several ways, but one method has
been most successful. A cloned gene is injected into the nucleus of a fertilized
egg cell. The cell is then implanted into the uterus of a foster mother
(Figure 12.4). The foreign DNA inserts into the DNA of the recipient early
enough in development to transform the progenitors of germ cells and
body cells alike. It is then inherited in succeeding generations in the same
way as all the other genes in the genome. Transgenic offspring and their
progeny are analyzed for the time and place of the transgene's expression
and for any abnormalities the transgene might create. Generally, the transgene includes important regulatory elements. Often, it is convenient to
attach the relevant regulatory sequences to a gene that encodes an easily
detectable protein rather than to the gene that is normally associated with
those regulatory sequences. This provides a simple way to assess the
organism's control mechanisms. In other instances, the effects of a specific
protein in a transgenic animal may be of more interest than the function
of a particular regulatory sequence. Here, the desired coding sequence can
be joined to a regulatory DNA sequence that will assure its expression,
rather than to the regulatory element normally associated with that coding
sequence. Molecular cloning makes the construction of such mixed genes
a straightforward task.

The study of transgenic mice can answer a variety of biological questions, some of which can be illustrated with an insulin gene. As mentioned
in Chapter 11, the expression of insulin genes is restricted to special
pancreatic cells because of the DNA segments immediately preceding the
insulin gene (Figure 11.2). When the insulin gene's regulatory region is
joined to the coding region of the SV40 oncogene—called large-T (see
Chapter 10)—and the recombinant DNA molecules are introduced into an
early mouse embryo, the genomes of all that embryo's descendants will
carry the integrated recombinant DNA. Because the coding sequence of
the large-T protein is now regulated by the promoter of the insulin gene,
large-T protein is synthesized exclusively in the insulin-producing cells of
the pancreas. Its oncogenic potential is manifest in the formation of tumors

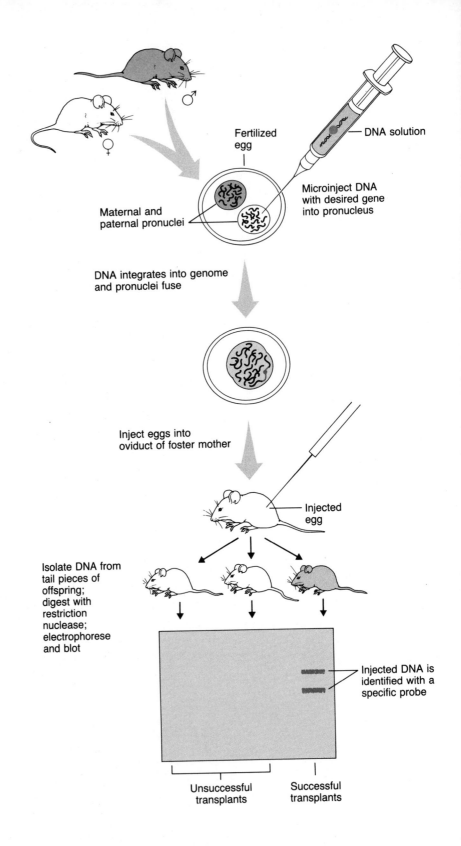

Fertilized egg

DNA solution

Maternal and paternal pronuclei

Microinject DNA with desired gene into pronucleus

DNA integrates into genome and pronuclei fuse

Inject eggs into oviduct of foster mother

Injected egg

Isolate DNA from tail pieces of offspring; digest with restriction nuclease; electrophorese and blot

Injected DNA is identified with a specific probe

Unsuccessful transplants

Successful transplants

Figure 12.4 (*facing page*)   Producing transgenic mammals. The colored mouse is a successful transgenic and will pass the new gene on to its offspring.

(called insulinomas) in these and only these cells. This experiment confirms the interpretation of experiments done with cultured cells; the insulin gene's regulatory sequences dictate that gene's expression only in a particular cell type.

A different result occurs if, instead of the insulin gene's promoter region, a control region from the gene encoding the enzyme elastase is used. Elastase is normally produced in pancreatic cells other than those that produce insulin. Here again, tumors occur in the transgenic mice. In this case, however, the tumors arise only in the pancreatic cells that make elastase rather than in those that produce insulin. Thus, tumors arise in two different cell types, depending on the regulatory sequence, even though the same large-T oncogene is used.

Such experiments provide a remarkable and widely applicable model system for the study of tumor formation and the development of precise therapeutic tools. For example, the introduction of a different oncogene, *myc*, attached to regulatory sequences from the genome of a virus that causes mammary tumors in mice, yields mammary tumors. This suggests that the specific association of this virus with mammary tumors is related to the expression of its genes preferentially in cells of the mammary gland. The experiment also confirms the characterization of the *myc* gene as an oncogene. Further, such transgenic mice can be used to study the efficacy of various drugs in inhibiting tumor growth.

Embryology

THE FEW experiments summarized thus far promise that the near future will solve long-standing riddles about complex biological phenomena. One such phenomenon is the development of a complex multicellular organism from a single fertilized egg cell. Although the process begins with one cell (or, at most, a few identical cells), many different kinds of cells must be produced during the progression to an adult organism. Molecular genetics has opened new windows on this marvelous process, long of singular interest to biologists. Of the many interesting experiments being done in plants and animals (including mammals), we can only mention a few.

During the cell divisions that follow fertilization, the newly produced individual cells develop different sizes, shapes, functions and potentials.

They also come to have specific spatial relationships with each other. All these events occur according to time-ordered patterns. In some species—mice for example—the cells formed in the first few cell divisions all have the same potential for further development into any part of the adult. In humans, this must also be true, at least for the first cell division, because each of the two cells can give rise to a complete human being in the formation of identical twins. As cell division proceeds, however, different cells become destined to give rise to only a limited number of adult cell types. When such "limited" cells divide, their daughter cells inherit their restricted fate.

Some cells become destined for other than embryonic tissue, such as the membranes that surround the fetus. Others will contribute to the formation of the gut lining, the nervous system, skin, muscles, blood vessels, or other tissues and organs. More restrictive specializations are programmed at subsequent cell divisions. Finally, the cells acquire the distinctive shapes and biochemical properties characteristic of the cells in each tissue of the newborn animal.

Eventually more than 200 cell types are distinguishable in vertebrates. A large body of evidence indicates that both timed and spatially distinctive patterns of gene expression are the underlying determinants of an organism's orderly development. Carefully controlled regulatory networks function in very early embryos. Some of these are sensitive to the nature of the neighboring cells. Others depend in complex ways on preceding patterns of gene expression. Thus, hierarchical expression patterns—that is, patterns in which the product of one gene triggers the expression of one or more other genes, and so on—dictate the orderly progress of successive and concomitant developmental stages. One aim of current work in developmental biology is to correlate the timed and three-dimensional changes that occur during early embryonic development with the expression patterns of specific genes. This approach has been particularly successful with fruit flies, largely because mutants are available with defects in early morphological development.

Anyone who has ever looked carefully knows that an insect's body has clearly demarcated segments. These segments, which are visible in both developing and adult flies, are normally associated with different anatomical features, such as eyes, antennae, wings, or legs (Figure 12.5). Classical genetics identified a group of fruit fly mutants in which particular anatomical features develop in the wrong segment. In one such mutant, for example, legs appear where antennae normally grow. Fly geneticists named the genes responsible for this type of change **homeotic genes** because the structures

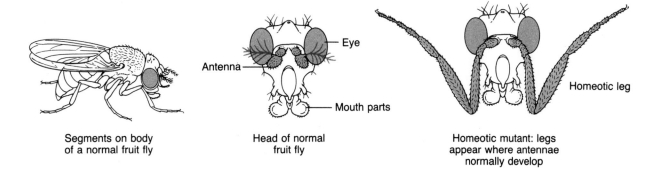

| Segments on body of a normal fruit fly | Head of normal fruit fly | Homeotic mutant: legs appear where antennae normally develop |

Figure 12.5   The mutation in the homeotic gene called *antennapedia* results in the growth of legs instead of antennae on the head of fruit flies.

that are made are like the normal structures (from the Greek *homoio-*, "similar") except for their location in the organism's body. The consequences of particular mutations in homeotic genes are generally localized to one or a few of the fly's segments. Such genes do not determine the number of segments formed but only the morphological development of given segments; mutations in other fly genes affect the number of segments.

Two enormous regions in the fly genome, each covering 250,000 base pairs or more, contain clusters of homeotic genes. Cell-specific transcription determines, at least in part, which particular homeotic genes will act in specific body segments. Thus, mRNAs for a homeotic gene are most abundant within the segment that is altered by mutations in that same gene. These observations demonstrate that morphological development is controlled by differential gene expression.

Genes that are expressed at specific times in development have been identified in organisms as diverse as corn, nematode worms, sea urchins, frogs, and mice. Libraries of cDNAs—that is, of copies of mRNAs—prepared from cells at particular developmental stages are used to identify those mRNAs that are being expressed at that stage. However, the experimental potential of these organisms depends on the availability of sophisticated genetic information. In the case of a tiny nematode worm, such information exists, and work is progressing rapidly. Not only has extensive classical genetic analysis and physical mapping of its genome been carried out, but the line of descent from the fertilized egg is known for every one of the 959 cells in the adult worm. It is possible to identify which genes are expressed and when, as particular cell lineages branch off in the course of the organism's development. Successful techniques for the introduction of cloned genes and artificial mutations into the germ cells of these worms are permitting even more precise investigations.

Among mammals, mice are the most promising animals for the analysis of developmental processes. Extensive classical genetic information exists, and many mouse mutations are known. Moreover, it is relatively easy to obtain cells from early developmental stages. From mice, and to some extent from other mammals, very early embryonic cells and cells from teratocarcinomas—tumors that consist of abnormally proliferating germ cells—are available and grow well in laboratory dishes. These systems provide promising new avenues for studying mouse embryology.

One of the new lines of investigation has its origins in the finding that the mouse genome contains genes that are very similar to some of the homeotic genes of fruit flies. The coding regions in many of the fruit fly's homeotic genes include almost identical DNA sequences that are 180 base pairs long—"**homeoboxes**"—that encode closely related polypeptide regions. These polypeptide regions impart to their proteins the ability to bind to certain DNA sequences that regulate transcription of many genes that are important in the fly's embryonic development. In fact, sequential production of these proteins is believed to create the cascade that guides the orderly events necessary for building a new organism. DNA sequences similar to those in fruit fly homeotic genes also occur in yeast, invertebrates, and vertebrates, including mammals. Moreover, mRNA molecules containing such homeo-box sequences are found in embryonic mouse cells soon after fertilization.

The sequential expression during the early stages of mammalian embryogenesis of a large number of mammalian genes that contain homeo boxes suggested that, in mammals as in flies, the family of proteins encoded by these genes influences the embryo's development. Very recent experiments using transgenic mice confirm this idea. Increases in the amount of a protein encoded by a particular mouse homeotic gene lead to localized abnormalities in the neck vertebrae of newborn mice. The increased level of protein resulted from the introduction of the homeotic gene as a transgene; the abnormal transgenic mice do not survive after birth. Thus, distantly related organisms, such as flies and mammals (and toads as well), utilize similar strategies for the establishment of the front-to-back organization of their bodies. For that reason, the wealth of information about fly development could lead to substantial progress in understanding mammalian development. Indeed, one of the fly's genes needed for normal wing development is known to be virtually identical with an oncogene that causes breast tumors in mice.

Making Mutations

ANOTHER experimental approach to understanding gene function in mammalian development takes advantage of the finding that transgenes may be inserted into genes that are important for normal development. We mentioned earlier that transgenes are useful for studying the regulatory signals that control the timing and tissue location of gene expression and for modifying mammalian organisms for experimental purposes (Figure 12.4). Because transgenes enter the genomes of recipient cells at unpredictable positions, they have also been useful for studying the effect of chromosomal location on the expression of genes—in this case, the transgene. But the random insertion of transgenes into a mouse's genome is also helpful for studying the function of the DNA region into which the transgene is inserted—the "target" site. This approach has been particularly helpful in identifying genes that are important in development.

Sometimes, the target site at which the transgene is inserted into the mouse genome will be the coding or regulatory region of an important gene. As a result, the gene at the target site will suffer a mutation. Its function may be blocked or modulated, or sometimes an altered protein is produced. The formation of a mutated gene at the target site may be quite independent of whether the transgene is expressed or not.

The disruption of genes by insertion has been a standard tool of bacterial, yeast, fruit fly, and corn genetics for many years. In these instances, mobile elements are used to disrupt genes. With mice, however, transgenes do the disrupting. Heritable mutations can also be produced by infection of very early mouse embryos with a retrovirus; in this case, if the retrovirus genome inserts into or near a gene, the gene's function is usually affected. For example, the gene's activity may be blocked, reduced, or enhanced; enhancement of cellular oncogenes frequently results in a cancer.

First-generation transgenic mice (that is, mice that develop from a treated fertilized egg cell) will have the transgene inserted into only one of a pair of homologous chromosomes (that is, in only one allele of the target site). By appropriate inbreeding, more heterozygous and some homozygous mice can be produced. Mice that are homozygous for the transgene are most likely to show an altered trait because both alleles are altered, and the consequences of the mutation may be observed. When interesting alterations are observed, the mutated gene at the target site can be cloned, even if it is completely unknown. This is because the inserted segment generally includes at least one DNA sequence that is foreign to the mouse

Transgenic   Normal

Figure 12.6 Fetal mouse with defective limb bones (*left*) as a result of the insertion of a transgene into a gene needed for normal limb development.

genome—the vector used for cloning the transgene. Thus, using probes that detect vector DNA sequences, it is possible to detect and recover from genomic libraries DNA clones that contain the altered gene; also, such clones frequently contain mouse sequences that surround the target gene.

In one example that illustrates how useful this method is, a mutation caused by the insertion of a transgene resulted in severe defects in the development of the limb bone of a mouse (Figure 12.6). The defects were the same as those observed more than 20 years ago in mice carrying a mutation that arose spontaneously. By chance, the target of transgene insertion proved to be an allele of the mutant gene detected earlier. That gene, and cDNAs representing mRNAs from that gene, have been cloned, and their sequences have been analyzed. Because similar genes have also been identified in the DNA of chickens and humans, it seems likely that this gene is essential for proper development in vertebrates. Studies of the expression of this gene and the nature of its protein product will shed light on this step in the developmental process. Other mutations caused by insertions are being identified in similar ways. Particularly important is the opportunity to identify genes that are essential in very early development. Mutations in such genes are likely to be lethal, leading to early abortion of the embryo. Nevertheless, there is a strong likelihood that studies of such mutations in mice will tell us a great deal about critical stages in human development.

Genetic Engineering of Plants

THE EXPERIMENTAL modification of plant genomes shares with the animal experiments the goal of understanding fundamental biological processes. With plants, the process of tissue differentiation and the mechanism by which light induces the expression of specific genes are important fundamental questions. Other research is motivated by a desire to improve the quality of agriculturally important crops. New and altered genes are introduced into single plant cells growing in laboratory dishes by means of special recombinant DNA vectors. Then, whole plants can be generated from the single cells, even though they are not germ cells, a process that does not work with animals.

Many experiments with plants depend on vectors derived from a special family of bacterial plasmids. This plasmid family is found, in nature, in the crown-gall bacterium, *Agrobacterium tumefaciens*. Together, the plasmids and

Normal plant   Transgenic plant

Figure 12.7 Caterpillars do not defoliate a tomato plant carrying a bacterial gene encoding a protein toxic to caterpillars (*right*). The gene came from *Bacillus thuringiensis*.

the bacterium constitute a natural mechanism for gene transfer to plants. Cells of *A. tumefaciens* cells bind to wounded plant tissue. Plasmids move from the bacteria into plant cells, and part of the plasmid DNA is inserted into the DNA of the plant cell. Some genes in the inserted segment of plasmid DNA are expressed in the plant cells. As a result, the plant cells form a tumor called a crown gall. Astonishingly, whole, healthy plants can be induced to grow from gall cells. This is the only known example of the natural transfer of genes between a bacterial plasmid and a eukaryotic genome. The system is adapted for plant genetic engineering by using recombinant DNA techniques to introduce "foreign" genes into the *A. tumefaciens* plasmid, and then into plants.

Whole plants that produce fertile seeds carrying functional foreign genes have been produced from cells transformed by these vectors. Genes that increase the resistance of crops to virus diseases or herbicides or that produce insecticidal compounds have already been introduced into crop plants (Figure 12.7). Many of these are now undergoing field trials. Other research is focusing on improving the nutritional quality of food crops.

Epilogue

MOLECULAR genetics has made a beginning; a wealth of detail about many biological systems is already available. But the successes do not amount to a complete or even a very profound understanding. On the contrary, current ignorance is vaster than current knowedge. Even in our high-technology world, nothing rivals the complexity and diversity of living things. No information system built by people approaches, in content, the amount of data encoded in genomes or, in complexity, the intricate regulatory networks that regulate gene expression. It is certain that there are in nature, remaining to be discovered, mechanisms and concepts that no one has yet even imagined.

What has been learned thus far—and summarized in this book—is analogous to having figured out the alphabet, the structure of some words, and a few rudiments of the grammar of a language that is yet to be deciphered. A few simple sentences now make sense to us. Studies of how the expression of genes is regulated may identify the grammatical rules. But great challenges lie before us. Deciphering this language is made more difficult by its many dialects—the different versions used in different biological systems. Yet at some future time we will know enough to read the language fluently. Then, the biological significance of DNA sequences will be recognizable to an educated eye—or an educated computer. For now, however, we have progressed only far enough to be able to identify important areas of ignorance. Certain of these concern longstanding questions, such as those about the details of development and differentiation, or the molecular basis of the mind. Others are new questions, raised by the very achievements of molecular genetics. And of course, we should be wary: some things that we now think we know may become less clear in the years to come, or even prove to be utterly wrong.

Certain of the questions likely to engage the energies of scientists in the next few years are already plain. Considerable progress has already been made in identifying the normal biochemical functions of certain oncogenes, although many remain mysterious; we do not yet know the critical

241

difference between the normal and the oncogenic forms. At this time, there is no adequate description of how a cellular or viral oncogene turns a normal cell into a tumor cell. How many cellular aberrations, other than oncogene expression, are required to yield a tumor cell? And, most importantly, can the oncogenic process be interrupted by precise molecular tools?

The immune system still holds many important secrets. For example, why do only 5 percent of the cells that enter the thymus from the bone marrow eventually emerge as functional T lymphocytes? How does the passage of those cells through the thymus educate the T lymphocytes to distinguish the molecules naturally present in the organism from those that are foreign? What has gone wrong in diseases in which the immune system fails to develop or fails to function normally? The urgency lent to such questions by the AIDS epidemic is stimulating a great deal of work in this area.

Genetic recombination was first described early in the twentieth century. At that time, homologous recombination in meiosis was the only kind known. Now, it is known that there are many different kinds of recombination and that they occur both in meiotic and in mitotic cells. Which type of recombination predominates in different organisms, and when (in development) and where (in what cell types)? Are there mechanisms to modulate recombination? Can we find ways to use a cell's recombination machinery to target genes to their correct positions for gene therapy rather than relying on their insertion into random positions? These questions now engage many investigators.

Because the activity of genes is influenced by their interactions with proteins, the structure of chromatin—the protein-bound form of chromosomal DNA—is another question that attracts attention. Can an understanding of the structures of such DNA–protein complexes help us to understand what "open, actively expressed" and "closed, silent" genomic regions are and how they work? Does the interaction between a regulatory protein and the DNA sequence it recognizes influence local chromatin structure and, thereby, the function of the genes in that region? How many different regulatory proteins exist? How is their synthesis activated so that gene function is modulated in specific ways?

The recently initiated project to map and sequence the entire human genome, an international enterprise, has major implications for research on these and other challenging questions. New insights into genome organization, new genes, new mechanisms for the regulation of gene expression, and new evolutionarily important processes are certain to be discovered.

The project, as envisioned, has a broad scope. Part of the effort will be devoted to the maps and nucleotide sequences of other genomes, such as those of yeast, worms, flies, and mice. Many of the genes in these organisms have human counterparts that are similar in nucleotide sequence and function, as any reader of this book knows. Therefore, ideas about the biological significance of human DNA sequence information will be testable in experimental systems. This is essential if the promise of the project is to be fully realized.

Molecular genetics epitomizes what is called the **reductionist** approach to biology. Single biological elements—such as a single gene—are examined in increasingly greater detail through a series of narrowly defined questions and experiments. The alternative, the **holistic** approach, assumes that little can be learned about the biology of organisms through reductionism. Rather, holistic biology insists that whole organisms should be studied within their environments in order to understand nature. Certainly, the vast majority of any organism's characteristics arise from complex interactions among many different proteins encoded by many genes. However, the expanding resources of molecular genetics—reverse genetics, transgenic organisms, and so forth—are now applicable to the study of complex multigene systems. Examples of this capability include the immune system, viruses, the construction of complex intracellular structures, such as ribosomes, and embryonic development. As the molecular designs of complex systems become comprehensible, the already arid discussions about reductionist versus holistic biology will become irrelevant. We can anticipate a continuum of biological understanding starting with individual genes, extending to the complex characteristics of complete organisms, and encompassing the intricate relations among organisms in ecosystems.

The fast pace of modern biology is driven by technology, by major challenges in health, agriculture, and industry, and, ultimately, by curiosity. Because we and our offspring are the products of functional genetic systems, our intellectual curiosity is buttressed by personal interest. This same combination of intellectual curiosity and personal relevance is responsible for the intense public interest in the recombinant DNA revolution and its product, genetic engineering. And it is why these achievements will identify twentieth-century biology as a great historical landmark. Thus far, the revolution has been a positive endeavor. Our understanding of ourselves and of other organisms has deepened. Important products—hormones, vaccines, and enzymes for research and commerce, for example—are being developed and produced. Harmless microorganisms are being modified so that they can be used to degrade oil in spills, detoxify wastes, and perform other beneficial environmental tasks. Still, there are disquieting aspects to the revolution. Care is required to ensure that no unexpected detrimental properties accompany the beneficial characteristics introduced into altered organisms. Deep thought and cautious analysis are required, if society is to avoid unwise use of the sophisticated diagnostic techniques afforded by the new methods, or of somatic-cell gene therapy, should it prove effective and practical. Even more profound are the questions that will arise if the techniques used to modify the germ lines of experimental mammals are ever considered for use in humans. The challenges are great, particularly because future research is inherently unpredictable in outcome.

It is sobering to us that almost nothing that is recorded in this book was known when we completed our formal educations more than 35 years ago. Our chief regret is that we are not likely to see the outcomes 35 years hence. We can be certain of only one thing: The major new concepts that will emerge from future research will be as unexpected in their time as were introns and mobile elements to the present generation of biologists. A changing perspective is a constant feature of the history of science, and molecular genetics will not be immune to that prospect.

Glossary

Note: Words set in *italics* are defined elsewhere in this glossary.

activator: A *protein* that enhances the *transcription* of a *gene* by binding either to a specific sequence of *nucleotides* or to other proteins bound to *DNA*.

allele: A version of a particular *gene*. *Eukaryotic cells* that contain pairs of *homologous chromosomes* contain two alleles for each gene in each body cell; the two alleles may be identical or different, but they occupy the same relative position on homologous chromosomes. *Bacteria* and eukaryotic *germ cells* (both *haploid*) have only one allele of any particular gene.

amino acid: Any one of the molecules that serve as building blocks for making a *polypeptide* (*protein*). Various arrangements of twenty different amino acids are commonly found in the proteins in all living organisms.

amplification: Increasing the number of copies of a *gene* or *DNA* sequence.

annealing: A process whereby two *complementary* single *polynucleotide* strands associate and form a double helix.

anonymous DNA segment: A *DNA* segment of unknown function.

antibody: Any one of numerous *proteins* of the immune system that circulate in the blood of vertebrates and bind substances foreign to the organism (*antigens*).

anticodon: The sequence of three *nucleotide* units on a *transfer RNA* (*tRNA*) molecule that is *complementary* to the *codon* for the *amino acid* specified by that tRNA.

antigen: A substance that elicits an antibody response.

bacteria: Free-living single *cells* that have no *nucleus* and are therefore classed as *prokaryotes*. Because their *genomes* contain only a single copy of each *gene*, they are termed *haploid*.

bacteriophage: A virus that infects *bacteria*; frequently called simply a *phage*.

base: The portion of a *DNA* or *RNA nucleotide* that is distinctive. The four bases in DNA are adenine (A), thymine (T), guanine (G), and cytosine (C). In RNA, uracil (U) replaces thymine.

base pair: The *complementary* pairs of *DNA* or *RNA bases* that stabilize double helices through *hydrogen bonds*: A with T (or U in RNA) and G with C.

B cell: A type of white blood *cell* (or *lymphocyte*) that synthesizes *antibodies* and secretes them into the blood. The "B" refers to the fact that, in birds, these cells develop in an organ known as the bursa. In mammals, which have no bursa, B cells derive from bone marrow and develop in a variety of other tissues.

bond: A link holding two atoms together (see also *chemical bond* and *hydrogen bond*).

catalyst: A substance that increases the rate of a chemical reaction between two other substances; no net change in the structure or amount of the catalyst occurs during the reaction, although it may be reversibly modified.

cell: The basic unit of life, capable of growing and multiplying. All living things are either single, independent cells or aggregates of cells.

cell membrane: A structure consisting of fats and *proteins* that encloses a *cell*. In animals cells, the cell membrane forms the outermost surface. Some other organisms, such as certain bacteria, yeast, and plants, have a cell wall consisting of complex sugars surrounding the cell membrane.

cell-surface receptors: *Proteins* in, on, or traversing the *cell membrane* that recognize and bind to specific molecules in the surrounding fluid. The binding may serve to transport molecules into the cell's interior or to signal the cell to respond in some way.

centromere: A specialized region on a *eukaryotic chromosome* that is necessary for the proper distribution of chromosomes to the daughter cells during cell division. Centromeres are often visible (in a microscope) as constrictions on mitotic chromosomes (see *mitosis*).

chemical bond: A connection holding two atoms together in a molecule. The link consists of one or more electrons orbiting around the nuclei of both atoms.

chloroplasts: Complex particles found in the *cytoplasm* of certain bacteria and plant *cells* that carry out photosynthesis, thereby capturing the sun's energy for the synthesis of biological molecules. The component of chloroplasts that absorbs the sun's radiation is chlorophyll, the pigment that gives green plants most of their color.

chromatin: A substance consisting of double-stranded *DNA* that is folded and condensed by virtue of being associated with special *proteins*.

chromosomes: Structures in *cell nuclei*, each of which contains a highly condensed *DNA* double helix associated with *proteins*.

clone: A population of organisms, *cells*, viruses, or *DNA* molecules that is derived from the *replication* of a single genetic progenitor.

cloning: The process of making a *clone*.

coding region: A stretch of *DNA* or *RNA* that contains a series of *codons* that can be *translated* into a *polypeptide*.

codon: Three consecutive *nucleotides* in a *DNA* or *RNA* chain that specify a particular *amino acid* or the beginning or end of a *coding region*.

complementary: *Bases* in *DNA* or *RNA* that can form *hydrogen bonds* with each other (*base pairing*) are said to be *complementary*. The complementary pairs are A and T (or U) and G and C. Thus, DNA or RNA chains can form double helices if the two chains contain stretches of complementary bases (e.g., 5'-CGCC-3' and 3'-GCGG-5').

complementary DNA (cDNA):   A *DNA* chain formed by copying an *RNA* chain using *reverse transcriptase*. The sequence of the cDNA is *complementary* to that of the RNA used as a *template*.

conjugation: The transfer of *DNA* from one *bacterial cell* to another across a *protein* bridge.

crossing-over: The interaction between two *homologous chromosomes* whereby portions of the chromosomes (and thus of their double-stranded *DNA*) are exchanged.

cytoplasm: The material within the *cell membrane* of an intact cell, excluding the *nucleus*. Nuclear and cytoplasmic contents are separated by the nuclear membrane.

denature: To alter the natural state of *nucleic acids* or *proteins*. Denaturation of *DNA* or *RNA* unwinds and separates the two strands in a double helix. Denaturation unfolds *proteins*.

diploid: A *cell* that has two copies of each of the *chromosomes* typical of the species—that is, has pairs of *homologous* chromosomes.

DNA (deoxyribonucleic acid):   Polymeric molecules composed of four different molecular units called deoxyribonucleotides (abbreviated A, G, C, and T) and containing genetic information (*genes*) encoded in the sequence of As, Gs, Cs, and Ts. Deoxyribonucleotides contain the sugar called deoxyribose.

DNA blotting: The transfer of *DNA* fragments from a gelatinous slab (used for electophoretic separation of the fragments) to an inert material, such as nitrocellulose, without disturbing the fragment pattern.

DNA fingerprint: The pattern of *DNA* fragments produced from a particular region of a *genome* by cleavage of DNA with a particular *restriction nuclease*, followed by separation of the fragments by *gel electrophoresis*. Depending on the particular DNA sequence being tested, the pattern may be unique for a single individual in a species.

DNA insert: A segment of *DNA* that is joined to a *recombinant DNA vector* for the purpose of *cloning*.

DNA ligase: An *enzyme* that joins the ends of *DNA* chains through the formation of strong *chemical bonds*.

DNA polymerase: An *enzyme* that elongates *DNA* chains by assembling individual *nucleotides* and joining them through the formation of strong *chemical bonds*. The precise order of the nucleotides is dictated by the *complementary* order of nucleotides in another DNA chain, the *template*.

DNA replication: The synthesis of two new *DNA* double helices, starting from one DNA double helix. Each chain in the first double helix serves as a *template* for the synthesis of a *complementary* chain.

enzyme: A *protein* or *RNA* chain that facilitates a specific chemical reaction. Virtually all intracellular chemical reactions are catalyzed by enzymes (see *catalyst*).

eukaryote: An organism whose *cells* contain a *nucleus*. The organism may be a single, free-living cell, such as a yeast, or a multicellular organism, such as a plant or an animal.

exon: A portion of a *gene* that is retained in mature, functional *RNA*; usually, but not always, containing sequences encoding a *polypeptide*.

expression vector: A *recombinant DNA vector* designed to permit *transcription* and *translation* of coding sequences contained in inserted *DNA* segments.

gel electrophoresis: A process whereby a mixture of molecules, such as *DNA* or *RNA* or *protein* molecules, is separated into its components, according to their size or electrical charge, on a slab of gelatinous material under the influence of an electric field.

gene: Initially, an abstract concept describing a unit of inherited information; now, a segment of *DNA* or *RNA* that constitutes a unit of inherited information.

gene conversion: A type of *recombination* between two nearly *homologous DNA* regions, whereby one of the DNA sequences is changed (converted) to match the sequence in the other; also called "sequence correction." This process can occur within one *chromosome* or between two chromosomes.

gene expression: The process whereby the information in a *gene* is used to produce a cellular component. Expression requires *transcription* of a gene's *DNA* sequence into *RNA* and, for most genes, *translation* of the RNA's sequence into a *polypeptide* (*protein*).

genetic code: The "dictionary" that relates specific *nucleotide* sequences in *RNA* or *DNA* to specific *amino acids* in a *protein*. Code words (*codons*) are a series of three consecutive nucleotides (such as AGG, GCA, and so on). Each codon specifies either an amino acid or the end of the coding sequence.

genetic map: A depiction of the linear order of *genes* along a *chromosome*.

genome: The totality of a *cell's* genetic information, including *genes* and other *DNA* sequences.

genome walking: Mapping the genomic *DNA* sequences in the vicinity of a *cloned* segment by using the ends of the cloned segment as *probes* to clone (usually from a *library*) the overlapping, contiguous DNA regions.

genomic blotting: *DNA blotting* in which the fragments from a *restriction nuclease* digest of total *DNA* from a particular organism's *cells* are visualized by *annealing* to a radioactive *probe* and exposure to x-ray film.

germ cells: *Cells* such as egg cells and sperm cells in animals and egg cells and pollen cells in seed plants. They have the property of giving rise to new, multicellular organisms after fertilization.

haploid: A *cell* that has only one copy of each of the *chromosomes* typical of the species.

heterozygous: A state in which two different versions of a particular *gene* (*alleles*) occur in a *diploid genome*.

homologous: Pairs of *chromosomes* or *DNA* segments that contain the same linear sequence of *genes* or *base pairs*, respectively. Homologous pairs of chromosomes or DNA molecules may differ in sequence to some extent, as when they carry different *alleles* of the same gene.

homozygous: A state in which the same version of a particular *gene* (that is, the same *allele*) occurs on both chromosomes in a *diploid genome*.

hormone: A relatively small molecule that serves as a signal to coordinate the activities of different *cells* and tissues of a multicellular organism.

hydrogen bond: A weak chemical interaction in which a hydrogen atom is shared between two oxygen or two nitrogen atoms or between an oxygen and a nitrogen atom. Hydrogen bonds are responsible for the base-pairing that holds together the two chains of a *DNA* double helix.

immune protein: An *antibody* or a *receptor* on the surface of a *cell* that can recognize and associate with a specific "foreign" molecule—that is, an *antigen*.

intervening sequence: See *intron*.

intron: A portion of a *gene* that does not encode any part of the final gene product and is removed (by *splicing*) from the mature, functional *RNA*.

library: A *recombinant DNA* library is a population of identical *vector* molecules, each of which carries a different *DNA insert*; altogether, the different inserts may represent an entire *genome* (for example, a human genomic library) or a collection of *messenger RNAs* isolated from a particular kind of *cell* and converted to the corresponding *complementary DNAs* (a cDNA library).

long terminal repeats (LTRs): Double-stranded *DNA* sequences, generally several hundred base pairs long, that are repeated at the two ends of *retrovirus* and *retrotransposon* DNA.

lymphocyte: A type of white blood *cell*; also found in other tissues, such as lymph nodes, spleen, and thymus. These cells are important components of the immune system.

macromolecule: A very large molecule that is polymeric—that is, made up of many similar basic building blocks. Examples are *DNA* (made up of *nucleotides*) and *proteins* (made up of *amino acids*).

meiosis: The process of *chromosome* sorting associated with *cell* divisions that leads to the formation of *germ cells*. After meiotic division, each daughter cell has only one chromosome from each pair of *homologous* chromosomes and is, thus, *haploid*.

membrane: A two-layered structure, composed of fatty material, that occurs within and surrounding all cells; most natural membranes have various *proteins*, such as *receptors* and *enzymes*, embedded in them.

messenger RNA (mRNA): A single *RNA* chain that can be decoded into a *protein*.

methyl: A group of atoms containing one carbon atom (C) *bonded* to three hydrogen atoms (H): —CH_3.

mitochondria: Complex structures found in the *cytoplasm* of virtually all *eukaryotic cells* that use oxygen to "burn" the chemical products derived from foodstuffs to produce energy for the chemical and mechanical processes carried out by cells; this process is referred to as respiration.

mitosis: The process of *chromosome* sorting associated with normal *cell* divisions. After mitotic division, each daughter cell ends up with both chromosomes of each *homologous* pair and is, thus, *diploid*.

mobile elements: *DNA* segments that can move from one place to another in a *genome*; also called *transposable elements* and "jumping genes."

molecular cloning: The *cloning* of a *DNA* molecule.

molecular genetics: The study of the molecular structures and chemical mechanisms of heredity.

molecular map: A depiction of the linear order of *DNA* segments along a *chromosome*.

mutant: A *gene*, or an organism carrying a gene, that has acquired a *mutation*.

mutation: An alteration in *DNA* structure. The alteration may give rise to an altered *gene* product or to unusual regulation of *gene expression*, which may, in turn, give rise to abnormal traits.

nucleic acid: A long polymeric molecule composed of single *nucleotide* units held together by strong *chemical bonds*. DNA and RNA chains are nucleic acids that differ in that their nucleotide units contain the sugars deoxyribose and ribose, respectively.

nucleolus: A small body inside the *nucleus* of *cells* within which *ribosomes* are assembled from *ribosomal RNAs* and *proteins*.

nucleotides: The individual building blocks of *DNA* and *RNA*.

nucleus: The sac within a eukaryotic *cell* (see *eukaryote*) that contains the *chromosomes*. It is separated from the *cytoplasm* by a nuclear *membrane*.

oncogene: An aberrant or misregulated *gene* that fosters tumor formation.

operator: A *DNA* sequence adjacent to a prokaryotic *gene* (see *prokaryote*) that permits a *repressor protein* to control *transcription* of that gene (and often of a group of consecutive genes as well).

origin of replication: The position on a *DNA* chain where *replication* begins.

pathogenic: Disease-causing.

peptide: A molecule formed by joining two or more *amino acids* by strong *chemical bonds* (see *peptide bonds*).

peptide bonds: The strong *chemical bonds* that join a carboxyl group to an amino group. Such bonds hold *amino acids* together in a *peptide* and in a *polypeptide* chain.

peptide messenger: A *peptide* that carries signals between the *cells* of a multicellular organism and thereby coordinates physiological activities; peptide messengers are therefore *hormones*.

phage: Nickname for a *bacteriophage*, a *virus* that infects *bacteria*.

plasmid: A circular double-helical *DNA* that can *replicate* independently of the *genome* within a *cell*.

polymerase chain reaction (PCR): A means for making many copies of a *DNA* segment without *cloning*.

polynucleotide: A synonym for *nucleic acid*.

polypeptide: A long *peptide*, generally containing 20 or more *amino acids*.

primary structure: The number and order of the different *nucleotides* in a *nucleic acid* chain or of the different *amino acids* in a *polypeptide*.

probe: A *DNA* or *RNA* segment tagged with a radioactive atom (or with some other easily detectable chemical group) used to detect *complementary nucleic acid* sequences by *annealing*.

prokaryotes: Single-celled organisms that lack a *nucleus*; *bacteria*.

promoter: A sequence of *nucleotides* on *DNA* that is required for the initiation of *transcription* by *RNA polymerase*.

protein: A folded *polypeptide* chain or several interacting folded polypeptides.

pseudogene: A copy of a *gene* that occurs elsewhere in a *genome* than in the normal position of the corresponding functional gene and that cannot be expressed because of alterations in regulatory *DNA* sequences, or coding DNA sequences, or both.

reading frame: A *nucleotide* sequence on *DNA* or *messenger RNA* that can be decoded into a *polypeptide*. Reading frames begin with a *start codon* (ATG in DNA or AUG in RNA) and end with a *stop codon* (TAA, TGA, TAG in DNA or UAA, UGA, UAG in RNA).

receptor proteins: *Proteins* on (or in) the *cell membrane* that recognize and bind to specific molecules in the extracellular environment.

recombinant DNA: A *DNA* molecule containing DNA segments from different sources—either biological or chemically synthesized—and formed by joining the diverse DNA segments together by laboratory procedures.

recombination: The natural process by which new associations are formed between segments of *chromosomes* (and thus between segments of *DNA*).

replication: The process in which *DNA* or *RNA* directs its own duplication.

repressor: A *protein* that inhibits *transcription* of a *gene* by binding to a specific sequence of *DNA nucleotides*—to an *operator*, for example.

restriction fragment length polymorphisms (RFLPs): Differences in the size of *DNA* fragments produced from the same region of the *genome* when the DNA of different individuals in a species is cleaved with a particular *restriction nuclease*.

restriction nuclease: An *enzyme* that breaks *DNA* molecules at a specific *nucleotide* sequence.

retrotransposon: A *mobile element* whose movement in a *genome* depends on its *transcription* followed by *reverse transcription* of the resulting *RNA*.

retrovirus: A *virus* with an *RNA genome* that reproduces by first copying its RNA in *DNA* through *reverse transcription*.

reverse transcriptase: The *enzyme* that copies an *RNA nucleotide* sequence into a *DNA nucleotide* sequence.

ribosomal RNAs (rRNAs): *RNA* chains that are an integral part of *ribosomes*.

ribosomes: Particles that assemble *amino acids* into *polypeptides* using *messenger RNA* as a *template*. They occur in *cytoplasm* and are composed of *RNA* and *protein*.

RNA (ribonucleic acid): Polymeric molecules composed of four different molecular units called ribonucleotides (abbreviated A, G, C, and U) and containing the sugar ribose.

RNA polymerase: An *enzyme* that initiates and elongates *RNA* chains by assembling individual *nucleotides* and joining them through strong *chemical bonds*. The precise order of the nucleotides is dictated by the *complementary* order of nucleotides in a *DNA* chain, the *template*.

shuttle vector: A *recombinant DNA vector* that can *replicate* in the *cells* of more than one species.

somatic cells: All the *cells* of the body—the soma— except the *germ cells*.

splicing: The process that removes *intron* sequences from *RNA* molecules and joins the *exons* into a functional RNA.

start codon: The codon ATG (in DNA) or AUG (in RNA) that marks the beginning of a reading frame and the place where translation begins.

stem cell: A relatively undifferentiated *cell* that can both perpetuate itself and give rise to one kind or several kinds of differentiated (that is, specialized) *cells*. For example, stem cells of the blood system, which are maintained in the bone marrow, give rise to both red and white blood cells.

stop codons: The *codons* in the *genetic code* that mark the end of a *reading frame* and the place where *translation* ends: TAA, TGA, and TAG in *DNA* or UAA, UGA, and UAG in *RNA*.

T cells: Several types of *lymphocyte* (white blood cell) that play important roles in the immune system. Their name comes from the fact that they pass through the thymus in the course of their development.

telomere: An end of a *chromosome* (and thus of the chromosome's *DNA* double helix).

template: A *DNA* or *RNA* chain, when it is used to direct the order of assembly of *nucleotides* into a new *polynucleotide* chain by *complementary base-pairing*.

transcription: The process by which an *RNA polymerase* synthesizes an *RNA* chain.

transcription factor: A *protein* that facilitates the *transcription* of a *gene* by *RNA polymerase*.

transcription unit: All those *DNA* segments associated with a *gene* that are *transcribed*: coding regions, introns, and certain other DNA sequences that may precede or follow the gene on the *chromosome*.

transduction: The introduction of foreign *DNA* into *bacteria* by a *phage*.

transfer RNAs (tRNAs): A set of small *RNA* chains, each one capable of being joined to a specific *amino acid* and having, in its *nucleotide* sequence, three consecutive nucleotides (the *anticodon*) that are *complementary* to the *genetic code* word (*codon*) for that amino acid.

tranformation: A permanent, heritable change in the properties of a *cell* caused either by the insertion of a new *DNA* sequence acquired from outside the cell or by *mutation* (as in the transformation of a normal cell into a tumor cell).

transgene: A *gene* that has been introduced experimentally into an organism's *genome* and is passed on to the offspring (*transgenic organism*).

transgenic organism: An animal or plant whose *genome* has been altered by the introduction of new *DNA* sequences in such a way that the organism's offspring will inherit the new sequences.

translation: The process by which the genetic code (*codons*) in a *messenger RNA* is "decoded" into a *polypeptide* with a specific sequence of *amino acids*. Translation takes place on *ribosomes*.

transposable elements: Another name for *mobile elements*.

vector: A *DNA* molecule that can be joined to a foreign DNA segment so that the *recombinant DNA* can then be introduced into a *cell* where it can be replicated.

viral oncogene: A *gene* on a virus *genome* that fosters tumor formation.

Recommended Reading

Prologue

Listed here are documents and books that relate to the events and issues described.

M. F. Singer and D. Söll, 1973. Guidelines for DNA hybrid molecules. *Science* 181:1114.

P. Berg, D. Baltimore, H. W. Boyer, S. N. Cohen, R. W. Davis, D. S. Hogness, D. Nathans, R. Roblin, J. D. Watson, S. Weissman, and N. D. Zinder, 1974. Potential biohazards of recombinant DNA molecules. *Science* 185:303.

P. Berg, D. Baltimore, S. Brenner, R. O. Roblin, and M. F. Singer, 1975. Asilomar conference on recombinant DNA molecules. *Science* 188:991–994.

Lord Ashby, 1976. *Report of the Working Party on the Experimental Manipulation of the Genetic Composition of Microorganisms.* London: Her Majesty's Stationery Office.

C. Grobstein, 1977. The recombinant DNA debate. *Scientific American* 237(1): 22–33.

J. Richards, ed., 1978. *Recombinant DNA: Science, Ethics and Politics.* New York: Academic Press.

C. Grobstein, 1979. *A Double Image of the Double Helix: The Recombinant DNA Debate.* San Francisco: W. H. Freeman.

General Readings:

Except for the last book (which provides a general introduction to cell biology), the following books are collections of articles, many of which are listed separately under specific chapters. All of these references are for the general reader.

F. M. Burnet, ed., 1976. *Immunology* (readings from *Scientific American*). San Francisco: W. H. Freeman.

D. Freifelder, ed., 1978. *Recombinant DNA* (readings from *Scientific American*). San Francisco: W. H. Freeman.

P. C. Hanawalt, ed., 1980. *Molecules to Living Cells* (readings from *Scientific American*). San Francisco: W. H. Freeman.

P. H. Abelson, ed., 1984. *Biotechnology and Biological Frontiers* (AAAS publication 84-8). Washington, D.C.: American Association for the Advancement of Science.

C. I. Davern, ed., 1985. *Genetics* (readings from *Scientific American*). New York: W. H. Freeman.

[Editors of *Scientific American*], 1985. *The Molecules of Life* (a *Scientific American* book; articles from the September 1985 issue). New York: W. H. Freeman.

D. E. Koshland, Jr., ed., 1986. *Biotechnology: The Renewable Frontier*. (AAAS publication 85-26). Washington, D.C.: American Association for the Advancement of Science.

C. de Duve, 1985. *A Guided Tour of the Living Cell.* New York: Scientific American Books.

Much of the material summarized in this book is presented in various textbooks in a detailed and rigorous manner. Some of these textbooks are listed here.

D. Friefelder, 1987. *Molecular Biology*, 2nd ed. Boston: Jones and Bartlett.

J. D. Watson, N. H. Hopkins, J. W. Roberts, J. A. Steitz, and A. M. Weiner, 1987. *Molecular Biology of the Gene*, 4th ed. Menlo Park, Calif.: Benjamin/Cummings.

J. D. Rawn, 1988. *Biochemistry*. Burlington, N. C.: Carolina Biological Supply.

L. Stryer, 1988. *Biochemistry*, 3rd ed. New York: W. H. Freeman.

B. Alberts, D. Bray, J. Lewis, M. Raff, K. Roberts, and J. D. Watson, 1989. *Molecular Biology of the Cell*, 2nd ed. New York: Garland.

J. Darnell, H. Lodish, and D. Baltimore, 1990. *Molecular Cell Biology*, 2nd ed. New York: Scientific American Books.

B. Lewin, 1990. *Genes*, 4th ed. New York: John Wiley and Sons.

M. Singer and P. Berg, 1990. *Genes & Genomes*. Mill Valley, Calif.: University Science Books.

Chapter 1

The historical development of genetics, nucleic acid biochemistry, and enzymology are described in the following books and articles.

N. W. Horowitz, 1956. The gene. *Scientific American* 196(4):78–85.

C. Stern and E. R. Sherwood, eds., 1966. *The Origin of Genetics: A Mendel Source Book*. San Francisco: W. H. Freeman.

A. E. Mirsky, 1968. The discovery of DNA. *Scientific American* 218(6):78–88.

H. S. Stubbe, 1972. *History of Genetics: From Prehistoric Times to the Discovery of Mendel's Laws* (translated by T. R. W. Waters). Cambridge, Mass.: MIT Press.

J. S. Fruton, 1972. *Molecules and Life*. New York: Wiley-Interscience.

R. Olby, 1974. *The Path to the Double Helix*. Seattle: University of Washington Press.

F. H. Portugal and J. S. Cohen, 1977. *A Century of DNA*. Cambridge, Mass.: MIT Press.

H. F. Judson, 1979. *The Eighth Day of Creation*. New York: Simon and Schuster.

D. S. Goodsell, 1991. Inside a living cell. *Trends in Biochemical Sciences* 16: 203–206.

Chapter 2

R. W. Holley, 1966. The nucleotide sequence of a nucleic acid. *Scientific American* 214(2):30–39.

A. Kornberg, 1968. The synthesis of DNA. *Scientific American* 219(4):64–78.

M. F. Perutz, 1978. Hemoglobin structure and respiratory transport. *Scientific American* 239(6):92–125.

W. R. Bauer, F. H. C. Crick, and J. H. White, 1980. Supercoiled DNA. *Scientific American* 243(1):118–133.

P. Howard-Flanders, 1981. Inducible repair of DNA. *Scientific American* 245(5): 72–103.

R. D. Kornberg and A. Klug, 1981. The nucleosome. *Scientific American* 244(2): 52–79.

J. C. Wang, 1982. DNA topoisomerases. *Scientific American* 247(1):94–109.

J. E. Darnell, Jr. 1985. RNA. *Scientific American* 253(4):68–78.

R. F. Doolittle, 1985. Proteins. *Scientific American* 253(4):88–99.

G. Felsenfeld, 1985. DNA. *Scientific American* 253(4):58–67.

F. W. Stahl, 1987. Genetic recombination. *Scientific American* 256(2):90–101.

H. Varmus, 1987. Reverse transcription. *Scientific American* 257(3):56–64.

[Editors of *Trends in Biochemical Sciences*], 1989. Structure and function of proteins. *Trends in Biochemical Sciences* (special issue) 14:243–312.

M. Pines, 1988. *The Structures of Life* (NIH Publication No. 88-2778). Bethesda, Md.: U.S. Department of Health and Human Services.

B. Cipra, 1990. Mathematics untwists the double helix. *Science* 247:913–915.

F. M. Richards, 1991. The protein folding problem. *Scientific American* 264(1): 54–63.

Chapter 3

J. D. Watson, 1963. The involvement of RNA in the synthesis of proteins. *Science* 140:17–26.

F. H. C. Crick, 1964. The genetic code. *Scientific American* 207(4):66–74.

A. Rich and S. H. Kim, 1978. The three-dimensional structure of transfer RNA. *Scientific American* 238(1):56–62.

J. A. Lake, 1981. The ribosome. *Scientific American* 245(2):84–97.

Chapter 4

The historical development of the concepts and techniques that led to the recombinant DNA method is described in the following references.

J. Cairns, G. S. Stent, and J. D. Watson, eds., 1966. *Phage and the Origins of Molecular Biology*. Cold Spring Harbor, N.Y.: Cold Spring Harbor Laboratory.

J. Cairns, 1966. The bacterial chromosome. *Scientific American* 214(1):36–44.

S. N. Cohen, 1975. The manipulation of genes. *Scientific American* 233(1):25–33.

D. Nathans, 1979. Restriction endonucleases, simian virus 40 and the new genetics. *Science* 206:903–910.

M. F. Singer, 1979. Introduction and historical background. In J. K. Setlow and A. Hollaender, eds., *Genetic Engineering*, vol. 1, pp. 1–13. New York: Plenum.

R. P. Novick, 1980. Plasmids. *Scientific American* 243(6):102–127.

J. D. Watson, M. Gilman, J. Witkowski, and M. Zoller, 1992. *Recombinant DNA*, 2nd ed. New York: Scientific American Books.

Chapter 5

F. Sanger, 1981. Determination of nucleotide sequences in DNA (Nobel lecture, 8 December 1980). *Bioscience Reports* 1:3–18.

M. D. Chilton, 1983. A vector for introducing new genes into plants. *Scientific American* 249(6):50–60.

B. H. Howard, 1983. Vectors for introducing genes into cells of higher eukaryotes. *Trends in Biochemical Sciences* 8:209–212.

M. H. Caruthers, 1985. Gene synthesis machines: DNA chemistry and its uses. *Science* 230:281–285.

R. F. Doolittle, 1987. *Of URFS and ORFS: A Primer on How to Analyze Derived Amino Acid Sequences*. Mill Valley, Calif.: University Science Books.

C. DiLisi, 1988. Computers in molecular biology: Current applications and emerging trends. *Science* 240:47–52.

K. B. Mullis, 1990. The unusual origin of the polymerase chain reaction. *Scientific American* 262(6):56–65.

Chapter 6

P. Chambon, 1981. Split genes. *Scientific American* 244(5):60–71.

J. E. Darnell, Jr., 1983. The processing of RNA. *Scientific American* 249(4):90–100.

T. Maniatis, S. Goodbourn, and J. A. Fischer, 1987. Regulation of inducible and tissue-specific gene expression. *Science* 236:1237–1244.

J. A. Steitz, 1988. "Snurps." *Scientific American* 258(6):56–63.

J. A. Witkowski, 1988. The discovery of split genes. *Trends in Biochemical Sciences* 13:110–113.

S. L. McKnight, 1991. Molecular zippers in gene regulation. *Scientific American* 264(4):54–64.

T. Beardsley, 1991. Smart genes. *Scientific American* 265(2):86–95.

Chapter 7

R. J. Britten and D. E. Kohne, 1970. Repeated segments of DNA. *Scientific American* 222(4):24–31.

T. Friedmann, 1971. Prenatal diagnosis of genetic disease. *Scientific American* 225(5):34–42.

R. M. Lawn and G. A. Vehar, 1986. The molecular genetics of hemophilia. *Scientific American* 254(3):48–65.

F. H. Ruddle, 1982. A new era in mammalian gene mapping: Somatic cell genetics and recombinant DNA methodologies. *Nature* 294:115–119.

L. Grivell, 1983. Mitochondrial DNA. *Scientific American* 248(3):78–89.

J. Stranbury, J. B. Wyngaarden, D. S. Fredrickson, J. L. Goldstein, and M. S. Brown, 1983. *The Metabolic Basis of Inherited Disease*, 5th ed. New York: McGraw-Hill.

C. M. Croce and G. Klein, 1985. Chromosome translocations and human cancer. *Scientific American* 252(3):54–60.

P. Gill, A. J. Jeffreys, and D. J. Werrett, 1985. Forensic application of DNA fingerprints. *Nature* 318:577–579.

V. A. McKusick, 1986. The gene map of *Homo sapiens*: Status and prospects. *Cold Spring Harbor Symposium* 51:15–27.

C. T. Caskey, 1987. Disease diagnosis by recombinant DNA methods. *Science* 236:1223–1228.

A. W. Murray and J. Szostak, 1987. Artificial chromosomes. *Scientific American* 257(5):62–68.

National Research Council, Board on Basic Biology, 1988. *Mapping and Sequencing the Human Genome*. Washington, D.C.: National Academy Press.

V. A. McKusick, 1988. *The Morbid Anatomy of the Human Genome: A Review of Gene Mapping in Clinical Medicine*. Bethesda, Md.: Howard Hughes Medical Institute.

R. White and J.-M. LaLouel, 1988. Chromosome mapping with DNA markers. *Scientific American* 258(2):40–48.

E. S. Lander, 1989. DNA fingerprinting on trial. *Nature* 339:501–505.

E. S. Gershon, M. Martinez, L. R. Goldin, and P. V. Gejman, 1990. Genetic mapping of common diseases: The challenges of manic-depressive illness and schizophrenia. *Trends in Genetics* 6:282–287.

J. Marx, 1990. Dissecting the complex diseases. *Science* 247:1540–1542.

P. J. Neufield and N. Colman, 1990. When science takes the witness stand. *Scientific American* 262(5):46–53.

J. C. Stephens, M. L. Cavanaugh, M. I. Gradie, M. L. Mador, and K. K. Kidd, 1990. Mapping the human genome: Current status. *Science* 250:237–244.

R. K. Moyzis, 1991. The human telomere. *Scientific American* 265(2):48–55.

B. J. Trask, 1991. Fluorescence *in situ* hybridization: Applications in cytogenetics and gene mapping. *Trends in Genetics* 7:149–154.

C. Wicking and B. Williamson, 1991. From linked marker to gene. *Trends in Genetics* 7:288–293

Chapter 8

G. J. V. Nossal, 1964. How cells make antibodies. *Scientific American* 211(6):106–115.

R. T. Schimke, 1980. Gene amplification and drug resistance. *Scientific American* 243(5): 60–69.

P. Leder, 1982. The genetics of antibody diversity. *Scientific American* 246(5): 102–115.

S. Tonegawa, 1985. The molecules of the immune system. *Scientific American* 253(4):122–131.

P. Marrack and J. Kappler, 1986. The T cell and its receptor. *Scientific American* 254(2):36–45.

G. L. Ada and G. Nossal, 1987. Clonal selection theory. *Scientific American* 257(2):62–69.

H. M. Grey, A. Sette, and S. Buus, 1989. How T cells see antigen. *Scientific American* 261(5):56–64.

J. Rennie, 1990. The body against itself. *Scientific American* 263(6):106–115.

J. Marx, 1991. Getting a jump on gene transfer in *Drosophila*. *Science* 253:1093.

Chapter 9

F. Ayala, 1978. The mechanisms of evolution. *Scientific American* 239(3):56–69.

M. Kimura, 1979. The neutral theory of molecular evolution. *Scientific American* 241(5):98–126.

W. Gilbert, 1981. DNA sequencing and gene structure (Nobel lecture, 8 December 1980). *Science* 214:1305–1312.

J. E. Darnell, Jr., 1985. RNA. *Scientific American* 253(4):68–78.

R. F. Doolittle, 1985. Proteins. *Scientific American* 253(4):88–99.

A. C. Wilson, 1985. The molecular basis of evolution. *Scientific American* 253(4):164–173.

C. Wills, 1989. *The Wisdom of the Genes*. New York: Basic Books.

M. H. Brown, 1990. *The Search for Eve*. New York: Harper and Row.

L. L. Cavalli-Sforza, 1991. Genes, peoples, and languages. *Scientific American* 265(5):104–110.

L. Roberts, 1991. A genetic survey of vanishing people. *Science* 252:1614–1617.

E. Culotta, 1991. How many genes had to change to produce corn? *Science* 252:1792–1793.

J. Cherfas, 1991. Ancient DNA: Still busy after death. *Science* 253:1354–1356.

M. Hoffman, 1991. Brave new (RNA) world. *Science* 254:379.

Chapter 10

J. C. Fiddes, 1977. The nucleotide sequence of a viral DNA. *Scientific American* 237(6):54–67.

J. M. Bishop, 1982. Oncogenes. *Scientific American* 246(3):80–92.

R. A. Weinberg, 1983. A molecular basis of cancer. *Scientific American* 249(5): 126–142.

T. Hunter, 1984. The proteins of oncogenes. *Scientific American* 251(2):70–79.

M. S. Brown and J. L. Goldstein, 1984. How LDL receptors influence cholesterol and atherosclerosis. *Scientific American* 251(5):58–66.

S. H. Snyder, 1985. The molecular basis of communication between cells. *Scientific American* 253(4):132–141.

J. M. Bishop, 1987. The molecular genetics of cancer. *Science* 235:305–311.

R. C. Gallo, 1987. The AIDS virus. *Scientific American* 256(1):46–56.

A. S. Fauci, 1988. The human immunodeficiency virus: Infectivity and mechanisms of pathogenesis. *Science* 239:617–622.

J. Marx, 1990. Oncogenes evoke new cancer therapies. *Science* 249:1376–1378.

J. Marx, 1990. Genetic defect identified in rare cancer syndrome. *Science* 250:1209.

R. A. Weinberg, 1990. The retinoblastoma gene and cell growth control. *Trends in Biochemical Sciences* 15:199–202.

J. Marx, 1991. Possible new colon cancer gene found. *Science* 251:1317.

B. Moss, 1991. Vaccinia virus: A tool for research and vaccine development. *Science* 252:1662–1667.

P. Tiollais and M-A. Buendia, 1991. Hepatitis B virus. *Scientific American* 264(4):116–123.

Chapter 11

S. Benzer, 1973. Genetic dissection of behavior. *Scientific American* 229(6):24–37.

W. Gilbert and L. Villa-Komaroff, 1980. Useful proteins from recombinant bacteria. *Scientific American* 243(4):74–95.

W. J. Gehring, 1985. The molecular basis of development. *Scientific American* 253(4):152–162.

R. L. Brinster and R. D. Palmiter, 1986. Introduction of genes into the germ line of animals. In *The Harvey Lectures*, Series 80, pp. 1–38. New York: Alan R. Liss.

R. M. Goodman, H. Hauptili, A. Crossway, and V. C. Knauf, 1987. Gene transfer in crop improvement. *Science* 236:48–54.

J. R. Knowles, 1987. Tinkering with enzymes: What are we learning? *Science* 236:1252–1258.

L. Stryer, 1987. The molecules of visual excitation. *Scientific American* 257(1):42–50.

R. A. Lerner and A. Tromontano, 1988. Catalytic antibodies. *Scientific American* 258(3):58–70.

J. Cherfas, 1990. Embryology gets down to the molecular level. *Science* 250:33–35.

D. A. Melton, 1991. Pattern formation during animal development. *Science* 252:234–241.

J. Nathans, 1991. The genes for color vision. *Scientific American* 260(2):42–49.

G. A. Strobel, 1991. Biological control of weeds. *Scientific American* 265(1):72–78.

Chapter 12

U.S. Congress, Office of Technology Assessment, 1984. *Human Gene Therapy: Background Paper.* Washington, D.C.: U.S. Government Printing Office.

E. M. DeRobertis, G. Oliver, and C. V. E. Wright, 1990. Homeobox genes and the vertebrate body plan. *Scientific American* 263(1):46–52.

J. Marx, 1990. Human brain disease recreated in mice. *Science* 250:1509–1510.

[Editors of *Science*], 1990. Plant Biology. *Science* (special issue) 250:881, 923–966.

J. Rennie, 1990. The body against itself. *Scientific American* 263(6):106–115.

I. M. Verma, 1990. Gene therapy. *Scientific American* 263(5):68–84.

A. S. Moffat, 1991. Transgenic animals may be down on the pharm. *Science* 254:35–36.

Epilogue

Listed here are several books and reports that consider the societal impacts of the new biology.

U.S. Congress, Office of Technology Assessment, 1981. *Impacts of Applied Genetics: Microorganisms, Plants and Animals.* Washington, D.C.: U.S. Government Printing Office.

D. J. Weatherall, 1982. *The New Genetics and Clinical Practice.* London: Nuffield Provincial Hospitals Trust.

President's Commission for the Study of Ethical Problems in Medicine and Biomedical and Behavioral Research, 1982. *Splicing Life: The Social and Ethical Issues of Genetic Engineering with Human Beings.* Washington, D.C.: U.S. Government Printing Office

J. G. Perpich, ed., 1986. *Biotechnology in Society: Private Initiatives and Public Oversight.* Elmsford, N.Y.: Pergamon Press.

National Research Council, Board on Basic Biology, 1987. *Introduction of Recombinant DNA-Engineered Organisms into the Environment: Key Issues.* Washington, D.C.: National Academy Press.

E. K. Nichols, 1988. *Human Gene Therapy.* Cambridge, Mass.: Harvard University Press.

U.S. Congress, Office of Technology Assessment, 1988. *Mapping Our Genes: Genome Projects: How Big? How Fast?* Washington, D.C.: U.S. Government Printing Office.

U.S. Congress, Office of Technology Assessment, 1988. *Field-Testing Engineered Organisms: Genetic and Ecological Issues.* Washington, D.C.: U.S. Government Printing Office

S. Olson, 1989. *Shaping the Future: Biology and Human Values.* Washington, D.C.: National Academy Press.

National Research Council, Committee on Scientific Evaluation of the Introduction of Genetically Modified Microorganisms and Plants into the Environment, 1989. *Field Testing Genetically Modified Organisms: Framework for Decisions.* Washington, D.C.: National Academy Press.

C. R. Cantor, 1990. Orchestrating the Human Genome Project. *Science* 248: 49–51.

J. D. Watson, 1990. The Human Genome Project: Past, present, and future. *Science* 248:44–49.

J. H. Barton, 1991. Patenting life. *Scientific American* 264(3):40–46.

B. D. Davis, ed., 1991. *The Genetic Revolution: Scientific Prospects and Public Perceptions.* Baltimore, Md.: The Johns Hopkins University Press.

P. L. Pearson, B. Maidak, M. Chipperfield, and R. Robbins, 1991. The Human Genome Project—Do databases reflect current progress? *Science* 254: 214–215.

Index

Citations of figures are indicated by *f* immediately after the page number.

Acknowledgments for Photographs

FRONTISPIECE:
Permission to reproduce the lithograph entitled *Head and Shape 1* (1964) was graciously granted by the artist Nathan Oliveira and Smith Anderson Gallery, Palo Alto, California. Joseph Quever provided the photograph from which the frontispiece was produced. The original lithograph (27 1/8 inches x 22 3/8 inches) was printed on Rives BFK at the Tamarind workshop in Los Angeles, California.

Figure 1.1, page 12:
Mammalian blood cells: From *Tissues and Organs: A Text-Atlas of Scanning Electron Microscopy*, by Richard G. Kessel and Randy H. Kardon. © 1979 by W. H. Freeman and Company.
Cells in a maple leaf: Courtesy of Judith Croxdale.
Figure 1.2, page 13:
Needle tip and bacterial cells enlarged: Dr. Tony Brain and David Parker, Science Photo Library, Photo Researchers, Inc.
Slice through a bacterial cell: CNRI, Science Photo Library, Photo Researchers, Inc.
Figure 1.3, page 15:
Light micrograph: Courtesy of T. C. Hsu.
Electron micrographic 3-D image: Courtesy of J. B. Rattner.
Electron micrograph of fixed chromosome: Courtesy of H. Ris.
Figure 1.6, page 20:
Location of bands: Courtesy of Uta Francke.
Banded human male chromosomes: Courtesy of Uta Francke.
Figure 1.7, page 21: Courtesy of T. A. Donlon.
Figure 1.8, page 22: Courtesy of Uta Francke.
Figure 1.10, page 26: Courtesy of B. John.
Figure 2.5, page 41: Courtesy of G. F. Bahr.
Figure 2.7, page 43:
a: Courtesy of J. Griffith
b: Courtesy of A. K. Kleinschmidt
c: Micrograph is by U. K. Laemmli, as published in D. W. Fawcett, *The Cell*, 2nd ed. (Philadelphia: W. B. Saunders, 1981).

Figure 2.8, page 44: Courtesy of L. Chow.
Figure 2.13, page 51: Atomic model courtesy of S. M. Kim. See S. M. Kim et al., *Science* 185 (1974).
Figure 2.18, page 57:
a: R. C. Williams, University of California/Biological Photo Services.
b: Courtesy of A. D. Kaiser.
Figure 3.5, page 66: Courtesy of M. Boublik.
Figure 4.13, page 101: Joe Angeles, Washington University, St. Louis, MO.
Figure 7.5, page 145: Courtesy of David C. Ward.
Figure 7.7, page 152: Based upon a photograph of the *ras* locus RFLPs provided by Ray White.
Figure 7.11, page 160: Adapted from an image provided by A. Jeffreys.
Figure 8.5, page 165: Courtesy of C. L. Hedley, from M. K. Bhattacharyya et al., *Cell 60* (1/12/90), cover.
Figure 10.1, pages 192–193:
a–g: Courtesy of F. A. Murphy.
h: Courtesy of J. E. Dahlberg.
Figure 10.2, page 193: From D. M. Salunke, D. L. D. Caspar, and R. L. Garcea, *Cell 46*, (9/12/86), cover.
Figure 10.5, page 201: From E. Norrby et al., *J. Virol. 6* (1970): 237.
Figure 11.2, page 215: From *Histology: A Text and Atlas*, Johannes A. G. Rhodin, © 1974 by Oxford University Press.
Figure 12.6, page 238: After R. P. Woychik et al., *Nature 318* (1985), p. 36.
Figure 12.7, page 239: Courtesy of H. Schneiderman and the Monsanto Company.

COLOR PLATES:
I: Courtesy of Drs. Elena Smirnova and Andrew S. Bajer, Department of Biology, University of Oregon.
II: Alfred Lammer, Phototake.
III: *Maize*: Runk and Schoenberger, Grant Heilman Photography, Inc. *Snapdragons*: Enrico Coen and Rosemary Carpenter, John Innes Institute.
IV: From D. M. Salunke, D. L. D. Caspar, and R. L. Garcea, *Cell 46*, (9/12/86), cover.

Paul Berg

Paul Berg was born in New York City, received his early education at Abraham Lincoln High School, completed his undergraduate education at Pennsylvania State University, and earned his Ph.D. in biochemistry from Western Reserve University in 1952. After obtaining further research training at the Institute of Cytophysiology in Copenhagen and at Washington University in St. Louis, he joined the faculty at Washington University. In 1959, he moved to Stanford University, where he is now Willson Professor of Biochemistry and Director of the Beckman Center for Molecular and Genetic Medicine, Stanford University School of Medicine.

Dr. Berg is a member of the U. S. National Academy of Sciences and its Institute of Medicine, the American Academy of Arts and Sciences, the American Philosophical Society, and a Foreign Fellow of the French Academy of Sciences and of the Royal Society (London). He is currently Chairman of the National Advisory Committee on the Human Genome Project. In 1980, he received the Albert Lasker Medical Research Award and the Nobel Prize in Chemistry for his studies of the biochemistry of nucleic acids, particularly recombinant DNA. He was awarded the National Medal of Science in 1983.

Dr. Berg's research uses biochemical and molecular genetic approaches for the analysis of eukaryotic gene expression and recombination. His work on the chemistry and biology of mammalian and human genomes provides basic knowledge for understanding, preventing, managing, and curing genetic diseases.

Maxine Singer

Maxine Singer was born in New York City, educated at Midwood High School, Swarthmore College, and Yale University, where she received her Ph.D. in biochemistry in 1957. She was a postdoctoral fellow and then Research Biochemist at the National Institute of Arthritis and Metabolic Diseases, National Institutes of Health (NIH), from 1956 to 1975. In 1975, she moved to the National Cancer Institute, NIH, and became Chief of its Laboratory of Biochemistry in 1979. Her research has been in nucleic acid chemistry and enzymology and in molecular genetics. In 1988, she became President of the Carnegie Institution of Washington and Scientist Emeritus at the NIH, where she continues to lead a small research group working on transposable elements in the human genome.

As a member of the U. S. National Academy of Sciences, Dr. Singer served as Chairman of the Editorial Board of the Proceedings of the National Academy of Sciences and has also served on the editorial boards of the Journal of Biological Chemistry and Science Magazine. She is a member of the Pontifical Academy of Sciences, the Institute of Medicine, the American Academy of Arts and Sciences, and the American Philosophical Society. She was a fellow (trustee) of the Yale Corporation (1975–1990), and is a member of the Governing Board of the Weizmann Institute of Science and a director of the Whitehead Institute and of Johnson & Johnson. In 1988, Dr. Singer received the Distinguished Presidential Rank Award, the highest honor given to a civil servant, and in 1992 she was awarded the National Medal of Science.